THE
MILL

THE MILL

Produced and Designed by
WILLIAM FOX

Photography by
BILL BROOKS

Written by
JANICE TYRWHITT

Illustrations by HELEN FOX

McClelland and Stewart

ISBN: 0-7710-3193-9

The Canadian Publishers
McClelland and Stewart Limited
25 Hollinger Road, Toronto M4B 3G2

PRINTED AND BOUND IN CANADA

Page 1: Gears from the turbine that once drove the McArthur Woollen Mill at Carleton Place, Ontario.

Page 2: The Jamestown Windmill in Rhode Island at sunset.

Page 4: Flour dust and cobwebs cloud the windows of the Elliot Mill in Williamsford, Ontario.

Page 6: A fashionable eighteenth-century chaise at Old Sturbridge Village, Massachusetts.

CONTENTS

1. THE POWER OF THE NEW WORLD/13
The Wonders and Workings of Mills/15

2. THE HERITAGE THEY BROUGHT/31
Putting Water to Work/33
Harnessing the Wind/38
The Grist Mill: Gears and Stones/41
Other Uses of Power/47

3. SETTLEMENTS ON THE SEACOAST/63
Mills of the Pioneers/65
Sowing the Seeds of Industry/74
Ties with the Old World/79

4. THE DISCOVERY OF INDEPENDENCE/91
Yankee Ingenuity/93
The Growth of the Regions/94
The Fortunes of New France/95
The Spirit of Rebellion/102

5. PATRIOTS AND LOYALISTS/115
Fighting for Independence/117
Loyalties in Conflict/124

6. TWO INDUSTRIOUS REVOLUTIONARIES/137
Oliver Evans: Inventor/139
Samuel Slater: Industrialist/150

7. MEN WITH NEW POWER/163
Energy and Exploitation/165
Two Canadian Mills/167
New Solutions to Old Problems/173
From Tubwheels to Turbines/178
The Wealth of the Forests/183
The New World in a New Age/186

8. VOICES OF THE PAST, VISIONS OF THE FUTURE/195
Living with Memories/197
Good Times and Bad/209
The Power of the Future/210

Bibliography/220
Acknowledgements/222
Index of Mills/223

*This book is dedicated
to those who know that
the future is buried
in the past.*

KEEPSAKE MILL

Over the borders, a sin without pardon,
    Breaking the branches and crawling below,
Out through the breach in the wall of the garden,
    Down by the banks of the river, we go.

Here is the mill with the humming of thunder,
    Here is the weir with the wonder of foam,
Here is the sluice with the race running under —
    Marvellous places, though handy to home!

Sounds of the village grow stiller and stiller,
    Stiller the note of the birds on the hill;
Dusty and dim are the eyes of the miller,
    Deaf are his ears with the moil of the mill.

Years may go by, and the wheel in the river
    Wheel as it wheels for us, children, to-day,
Wheel and keep roaring and foaming for ever
    Long after all of the boys are away.

Home from the Indies and home from the ocean,
    Heroes and soldiers we all shall come home;
Still we shall find the old mill wheel in motion,
    Turning and churning that river to foam.

You with the bean that I gave when we quarreled,
    I with your marble of Saturday last,
Honoured and old and all gaily apparelled,
    Here we shall meet and remember the past.

Robert Louis Stevenson

Eplett's Mill in Coldwater, Ontario, was built in 1833 to serve the Indians settled by the government on a reserve between Coldwater and Orillia. The original building with its clerestory windows looks more like a textile mill than a grist mill. Now, with additions sprouting in all directions, at least four different window patterns, and faded red board-and-batten plastered with advertisements, it has become an amalgam of history.

1

THE POWER OF
THE NEW WORLD

Helen Fox

14

1

THE WONDERS AND WORKINGS OF MILLS

Our celebration of mills must begin with an acknowledgement of their beauty. Towering against the sky on high ground, a windmill is as magnificent as a square-rigged ship in full sail. A water wheel turning in a stream delights eye and ear like the surge and slow recession of waves on a shore. Worn cogs in immense wooden gear wheels mesh with a cumbersome elegance. In an old grist mill, every surface is wood worn to satin by the gentle abrasion of flour.

Yet anyone drawn to mills quickly recognizes that mills, like people, are infinitely various, and as interesting. Each has its own oddities, products of the millwright's skill or bad judgement, plus a measure of blind chance. In every mill, some parts work well and others have quirks that irritate the miller every time he sets his hand to them. Some mills have dropped out of history because they *never* worked and had to be abandoned or rebuilt. The most handsome are not always the most efficient. Long before architects, millers knew that form follows function, and their buildings took curious forms. Early windmills, wooden boxes perched on posts that allowed them to spin completely round with their sails to the wind, look ludicrous but they fed families. Most sawmills are stark sheds, often untidy, deafening in operation, but wonderfully productive.

The attraction of mills is elemental. They harness the natural elements of wind and water to turn raw materials into usable form. Because our most basic need is food, mills that ground grain were the first to make the crucial advance that substituted natural power for human labour. Without crushing or grinding, grain is almost inedible. The finest flour comes from the hardest grain, wheat. Its starchy inner kernel and oily cell of vitamin-rich germ are encased in a brittle cellulose shell of bran that must be crushed or, more efficiently, sheared off like the peel of an apple before the kernel can be ground to flour. Bran is nutritious, but until modern processing made it digestible it was used mainly to fatten pigs.

In ancient times, making flour was a full-time job. Even hand querns, the small grinding machines used aboard ships and in

◀ *Robert E. Lee's grist mill seems a remarkably gentle possession for a man of war. The mill at Lee's birthplace, Stratford Hall Plantation in Virginia, is thought to have been built about a century and a half before the Civil War. Reconstructed in 1939 with 200-year-old machinery from a Maryland mill, the mill's two run of stones, French burrs and granite, have been driven since 1958 by an eighteen-foot overshot wheel.*

remote districts up to our own century, were tedious to turn. Some pioneers brought them to America. A Nova Scotian wrote, "Grinding on the hand-mill was so laborious that it was let alone until necessity impelled to it. This was the occasion of saving much wheat, for many a meal was made without bread on account of the trouble of grinding."

Mills emancipated man from repetitive, time-consuming, backbreaking physical work by providing power from a source outside his own body. The recorded history of power begins with an account of a watermill written by the Roman architect Vitruvius about 15 B.C.: it used a vertical water wheel to transmit the energy of water through a series of gears to drive a pair of horizontal millstones. With some refinements, watermills have been working the same way ever since. Vitruvius' description is matter-of-fact. Its implications are astonishing. The idea of using water power to do work that had nothing to do with water – grinding grain – was entirely new. The mill had to be set in motion and controlled by the miller, but its production was no longer limited by the strength of his muscles. Mills were our first machines, and for the next two thousand years they played a central and continuing part in the developing use of power.

Now that they have been superseded by factories using mass-produced electricity, the wind and watermills of yesterday are in danger of falling victim to romantic legend. Ruined or restored, some mills seem such charming anachronisms that we forget how essential they once were to the life of our communities and the expanding economy of North America.

The first European immigrants came to North America from countries that offered little opportunity for the poor and unskilled. Political power was concentrated in the hands of landowners, courtiers, merchants, and manufacturers. In the New World, the settlers found a land rich in a different kind of power: wind and, more important, water. Leaping falls, spring-fed streams, ocean tides, and the stiff breezes of the Atlantic seaboard and the St. Lawrence shore could be harnessed to save time and work. Within a generation our ancestors re-created the development of milling. First they reverted to pestles and mortars made from logs and hollowed tree trunks, hand querns, sawpits, and mills worked by oxen or horses, and the use of these crude devices was carried on simultaneously with the building of wind and watermills in every village big enough to support the communal production of ground grain or sawn lumber.

Almost always, water and wind were put to use by a man with

Surgeon and sculptor Tait McKenzie from Almonte, Ontario, bought this mill near his hometown and renamed it the Mill of Kintail. The runner stone mounted in the foreground and the pieced French burrstone set in the footpath behind it are grindstones proclaiming that the building was originally a grist mill. McKenzie remodelled it, enlarging the windows to light his studio. Since his death in 1938, it has been preserved as a museum displaying his art.

Primitive rubbing stones.

A hand quern.

In North America, settlers suited their mills to their power sources. A smock mill (1) catches the wind on flat coast-land. Below it, a tidal mill (2) is driven by water from the millpond behind, trapped at high tide and gradually released as the pond empties. Directly above the tidal mill, an overshot water wheel (3) is fed through a flume from a mountain lake. Beside it, a flume from the same source pipes water to a turbine (4). Below, water from the millpond behind powers a sawmill's tubwheel (5). At top right, a waterfall turns an undershot wheel (6). At bottom right (7), a mill has its water wheel housed inside to protect it from ice in cold northern regions.

a little more initiative than his neighbours. As they might have put it, he could see further through a millstone than most. The meeting place of agriculture and industry, mills were the hub round which the life of each community revolved. The miller sensed the surge and ebb of economic tides and sometimes used his knowledge to intervene in critical events. Often, as the colonies consolidated, he rose to political prominence.

"Mills are a kind of property that yields a steady income," wrote Brissot de Warville in 1788, but the French revolutionary travelling the States was so fired with enthusiasm for independence that he saw only success. Milling could be far from profitable. Perhaps typical was a miller of Providence, Rhode Island, who died in 1682 leaving assets worth £90, of which £40 was the value of his mill. Some millers moonlighted as farmers, carpenters, blacksmiths, or fishermen. The right to sell fish from the mill-stream was a source of extra income that often went with the mill. From agreements of eighteenth-century apprentices, which specified that the lad "should not be required to eat salmon more than twice a week," we can guess that the miller's family sometimes depended on a diet of bread and fish. The fish fed the mill cats, too; every grist mill had its cat to catch mice, rats, and birds who robbed sacks of grain.

Customers were fickle. James Barry of Pictou County, Nova Scotia, who kept a crotchety illustrated journal of his mill on Six Mile Brook, complained in 1850, "People are flocking to me from all quarters with grain that I can grind for nothing. Of course they all tell bad of the Mill they left and they will soon do that with myself."

The miller knew the hazards of his trade, and its traditions: the cross scratched on the millstone to ask God's protection, the rule that work stopped at midnight Saturday, the duty to set windmill sails as guides for mariners in craft as perilous as his, the "unlucky" grindstones set as grave markers for millers killed by machinery that seemed to have a life of its own. James Barry recorded, "David McKay opened up the sawmill today. He did not intend to saw any, but of its own accord it sawed 2 feet into a log, just to let him see what it could do."

Fire was the calamity every miller dreaded. Working long hours in mills lit by candles and oil lamps, crammed with combustibles, some millers chewed tobacco but few smoked. "Burned out" is the recurring theme in mill records. "Phoenix" is the common mill name that testifies to the determination to rebuild. Most vulnerable of all were gunpowder mills. When a powder mill on Twelve Mile Creek near Waterdown, Ontario, exploded in 1884, the dead were laid out on the floor of the neighbouring Dakota watermill where Len Pegg still grinds oats.

The archives of state and provincial assemblies are packed with petitions from millers begging small cash bounties to rebuild after fires and floods. In 1828, Samuel McKeen of Mabou, Nova Scotia, asked the assembly to help restore his corn and sawmills: "In consequence of a heavy fall of rain, which carried with it a previous heavy fall of snow, the River of Mabou was swollen to a height such as was never before witnessed by the oldest person in the settlement and rushed down from the adjacent highlands with such impetuosity, that it swept along with it and completely destroyed the said Mills without leaving a vestige behind – that your Petitioner having thus in one hour been deprived of the fruits of many years labour – and not being possessed of sufficient means to re-establish himself, without some pecuniary aid – and as the loss of those Mills must be looked on as a Public calamity, and their re-erection highly necessary for the prosperity of the District.... He has been induced to Pray that your Honourable House will extend its bounty to him."

Early grist mills were usually the first granted water privileges, but later settlers built sawmills, then plank grist mills. Soon they were running multi-purpose operations from a single power source under one roof or in stands of mills. When several millers used the same river, disputes over water rights dragged on for years, and a burst dam upstream could destroy all the mills below.

Millers faced such disasters with remarkable resilience, and the fact that their sons and grandsons often carried on their work testifies to the satisfaction they found in it. Dugald Munn, advertising in *The Daily Examiner* in Charlottetown, Prince Edward Island, in 1888, wrote with exhilaration not wholly commercial: "The large trade done by Mr. Munn causes quite a 'hum' at the Roseneath Mills. A handsome and commodious new dwelling house has lately been built by Mr. Munn near the mill, and is an evidence that enterprize and industry are not without their reward even in Prince Edward Island."

Millponds were village playgrounds, swimming and fishing holes in summer, and skating rinks in winter. In 1817 John Jeffries from Somerset built a grist mill on the Piccadilly Brook at Sussex Corner, New Brunswick, where his seven sons worked three shifts round the clock. Now derelict, probably the oldest mill still standing in New Brunswick, it stayed in the family until 1952, and John's great-grandson Cecil Jeffries remembers when the children used to skate on the pond.

Landscape painters were obsessed by mills. John Constable, who had worked in his father's mills in Suffolk, kept a wooden model of a post mill to check the harmony and accuracy of his sketches. Homer Watson, whom Oscar Wilde called "the Constable of Canada," was the son of a woollen miller in Doon, Ontario. In his teens he worked in his grandfather's sawmill. Later, when it had been shut down, he painted it. Though he was only twenty-five and had never sold a picture, he sent *The Pioneer Mill* to the first exhibition of the Royal Canadian Academy in 1880. To his delight, the Governor General's wife Princess Louise bought it for $300 and gave it to her mother, Queen Victoria, who promptly ordered another picture by the man who became the most celebrated Canadian painter of his generation.

A detail from The Pioneer Mill *by* ▶
Homer Watson.

Dedham Mill, *painted by John Constable in 1820, was his father's watermill.*

Today, Andrew Wyeth lives in the restored stone millhouse beside his 1711 grist mill in the Pennsylvania hills.

Naturally enough, most country watermills are idyllically set beside brooks, their ponds hung with willows, alders, and flaming jewelweed. With his brother Reginald, Edward MacAusland of Bloomfield, Prince Edward Island, runs the woollen mill inherited from his father on the site of his grandfather's grist and sawmill. Busy making blankets for sale from Newfoundland to British Columbia, he still enjoys his surroundings. He writes, "At my front door is the millpond, a pond of beauty especially in the early hours of the morning. It is to me most interesting to see the industrious beaver, the old mother duck with her clutch of ducklings and the odd muskrat swimming about in the late evening."

These gentle places offer moments of tranquillity to anyone who seeks them out. Watermills are easy to find. They nestle into the landscape; abandoned, they disintegrate into a rubble of stone and rotting timber that in time returns to the river that gave them life. Look for a dip in the road, a hollow where a stream has been dammed to provide a head of water that transmits power from wheel or turbine to grindstones, saw, or machinery that cards, spins, and weaves yarn. Apparently simple, this process evolved through centuries, and was used without drastic change for a millennium.

Windmills are a different breed, conspicuous on flat land or hillsides. Toppled by a gale, they fall like scarecrows. A working windmill is a paradox. The sails are tuned to the rhythm of the wind, yet there is a maniac quality in the movement, the form, the very concept that brought them into being. Whose imagination first recognized that the force that swept ships out to sea could be used for an entirely different purpose on land? It was a fanciful notion, but it worked.

When you visit restored or running mills, take advantage of the chance to explore them. They reveal only gradually the complexity of their inner workings, their strengths and stubborn flaws, the variety that gives each its peculiar distinction. Their machinery illustrates the first tremendous advance in the use of natural power, and the gradual evolution of mechanical engineering. Our ancestors rejoiced in the mills that freed them from manual labour. Meanwhile, the millers were discovering principles of practical science to which we owe the leisure time that we take for granted, but which they would have considered truly marvellous.

Each sail of the Prescott Farm ▶ Windmill at Middletown, Rhode Island, carries 180 square feet of canvas. Here the canvas is furled. To set his sails the miller climbs the light, flexible sail bars like a ladder.

THE WINDMILL

1

2

Windmills Along the Shore

1. Two centuries ago the fieldstone mills of Quebec were a common sight along the St. Lawrence shore. Simply by surviving since 1772, the Moulin Desgagné on Ile-aux-Coudres has become a kind of natural marvel like Percé Rock or Montmorency Falls.

2. A verdant Long Island park is a picturesque but impractical setting for the Bridgehampton smock mill, surrounded by trees that rob wind from its sails.

3. Untarnished by time, this reproduction of the corn mill on Spocott Plantation at Lloyds, Maryland, looks as tidy as its original must have been when it was built in 1850 – very late for a post mill, the earliest type of windmill. The red cartwheel in the foreground is attached to a pole projecting from the back of the building. The entire body of a post mill pivots on its strong centre post. When the wind changes, the miller uses the tail pole as a giant handle to turn the mill body so that the sails face into the wind. The cartwheel at the end of the pole makes his job much easier.

4. The smock mill at Prescott Farm, one of the biggest ever built in the United States, weighs eighty-five tons. More than a third of this weight is moving machinery – cap, sails, windshaft and gears. The 36-foot windshaft alone weighs three thousand pounds.

5. Six steeply angled sails lend a rakish air to the mill in the Jardin Zoologique de Québec at Orsainville, a copy of a seventeenth-century tower mill.

6. Like most North American smock mills, the Pantigo Mill in East Hampton, Long Island, is clad not in clapboard but in shingles, more weather resistant and easier to patch after Atlantic gales.

7. New England windmills were peripatetic creatures, often moved to more useful locations. Built at Plymouth, Massachusetts, in 1793, and dismantled and hauled across the frozen bay a hundred years later, the smock mill at Eastham is the oldest windmill on Cape Cod.

3

4

5

6

7

Turned to the Wind

1. At Prescott Farm the miller, standing on the ground, rotates the cap of his mill by hauling on the endless chain that turns this winch. The winch is geared to the curb on which the cap revolves.

2. The weathervane of the mill at Ile-aux-Coudres reminds us that New France depended on the beaver trade. The thin board sails are characteristic of Quebec windmills.

3. The cap of the Eastham Windmill on Cape Cod is so heavy that it was turned by a horse or ox when the mill was in daily use. At the end of the tail pole is a metal rod, originally the first in a jointed series. The lower rods have been removed because they offered too great a temptation to souvenir hunters. Guy wires stretching from cap to tail pole strengthen the system.

4. The cap of the mill at Bridgehampton, New York, is turned by the fantail at the back. The whale weathervane affirms the affinity of millers and fishermen in seaport settlements.

5. In 1970 the mill at Jamestown, Rhode Island, was fitted with a new windshaft from the trunk of a long-leaf yellow South Carolina pine. Four years later a hurricane snapped off all four sails, and the windshaft was replaced with a red oak timber from Connecticut. The star carving is copied from the earlier windshaft.

6. The centre post of the Spocott Windmill at Lloyds, Maryland, was hand hewn from a 200-year-old white oak, almost a century older than the original mill. The post is braced by quarter-bars resting on brick piers.

7. A close look at the tail pole of the James Corwith smock mill at Watermill, Long Island. The weathered timber is attached to the axle of a cartwheel with wooden spokes and an iron rim. The wheel can revolve full circle on the grassy turf around the mill, so that by pushing the tail pole either to the left or to right the miller can turn the cap of his mill so that his sails face the wind.

1

2

3

4

5

6

7

27

Sails on the Seacoast

The windmills of Long Island, Rhode Island, and Cape Cod were often built by ships' carpenters or captains skilled in shaping and planing curved ribs and balancing structures to work with the weather instead of fighting it. On the coast, the set of windmill sails served as weathervanes for homecoming sailors and fishermen.

Lacking the rich ornamentation of some mills in the Netherlands and other countries on the Continent, the seaboard smock mills and sturdy Quebec tower mills acquired their own kind of beauty in the texture of weathered shingles or mellow masonry. New Englanders still insist that grain ground in a windmill has the salt taste of the sea and a finer texture and flavour because the slow, intermittent power of the wind prevents overheating between grindstones.

Like old wooden yachts, windmills are highly susceptible to weather damage and extravagantly expensive to maintain. The few that remain in good condition owe their survival to historical associations and foundations. The octagonal smock mill at Jamestown, Rhode Island, is typical. In 1787 the town wardens authorized payment of £100 to build it. The last miller, Jesse Tefft, stopped grinding in 1896 and the mill was abandoned, its sails shattered by storms, its furnishings filched for firewood. From 1912 to 1961 the Jamestown Historical Society spent more than $15,000 on repairs. In 1970, the Society raised $27,500 for a full restoration. When it was wrecked by a hurricane in 1974, the cost of repairs was covered by insurance.

The windmill at Chatham, Massachusetts, built in 1797 for Colonel Benjamin Godfrey, was moved to Chase Park in 1956. On its earlier site on Mill Hill it served as a landmark for sailors off the southeast tip of Cape Cod. With a high wind it could grind a bag of corn into meal in ten minutes. With a light wind, it took all day.

Three photographs of the windmill at Prescott Farm, Rhode Island, show how power from the sails is transmitted through gears to drive the grindstones. Above, the great brake wheel at the right is mounted on the windshaft that turns with the sails. The brake wheel meshes with the cast-iron wallower in the foreground, rotating its axle, a wooden spindle.

Below left: The wooden spindle from the wallower extends down through the floor to turn the gear at its base. This middle gear drives the two side gears in foreground and background, and each of these 600-pound "stone-nuts" turns the runner stone of a pair of millstones.

Below right: Poised above the millstones is the hopper that feeds them grain. At Prescott Farm the hopper is filled with Rhode Island whitecap flint corn, back-bred from modern hybrid corn to the original eight-row strain the first settlers used for johnnycake.

2

THE HERITAGE
THEY BROUGHT

2

PUTTING WATER TO WORK

The first Europeans who settled in the New World knew milling as a highly developed craft in their homelands. England alone has had 27,000 watermills and 10,000 windmills. From France, the mill designs and rights of landowners and religious orders were transferred to the Canadian colonies. In the seventeenth century, when Dutch trading ships circled the world, windmills of the Netherlands had reached a peak of mechanical perfection. Germans and Scandinavians were skilled in sawmill construction. Although early North American mills were simple, their builders drew on memories of sophisticated models.

The first generation of immigrants ground corn by methods literally prehistoric. The cavemen beat wild grain with chunks of stone to crush the hard bran shell. By the time men began cultivating cereals in the Middle East, centuries before the birth of Christ, they had developed mortars and pestles, hollowed stones to hold the grain and stone rods to pound it. In northern timberlands, mortars were scooped from tree trunks and pestles whittled from logs.

By 3000 B.C. Egyptians were using the saddle quern, an oval or oblong stone on which grain was rubbed with a flat-bottomed stone rod. Kneeling at one end of the slanted saddlestone, the wife or maidservant sheared off the bran and ground the kernel into meal. The Greeks made improvements that have been used in milling ever since. They cut grooves in the working surfaces of the stones to make it easier to remove the bran without crushing it, and they enlarged the upper stone into a bowl or box with a slot at the bottom through which grain could be fed continuously to the grinding surface.

True millstones evolved from the rotary quern, a pair of roughly circular, flat stones bored with a central hole to mount a pivot. A handle was set near the edge of the upper stone and was used to swing the stone back and forth in an arc. After centuries, some brilliant innovator realized the top stone could be turned all the way round in a complete circle. Thus continuous motion was introduced to milling.

◀ *The village of Roblin's Mills in the Quinte Loyalist settlement at Ameliasburgh grew round the five-storey stone flour mill built in 1842 by Owen Roblin. The mill, dismantled and rebuilt in 1963, can be visited at Black Creek Pioneer Village in Toronto. The man who made it is best discovered in Al Purdy's many-layered narrative poem,* In Search of Owen Roblin.

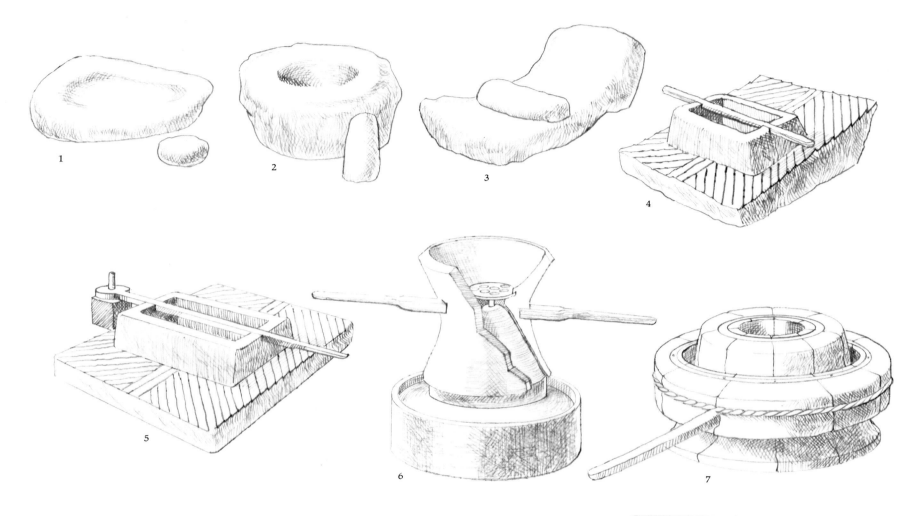

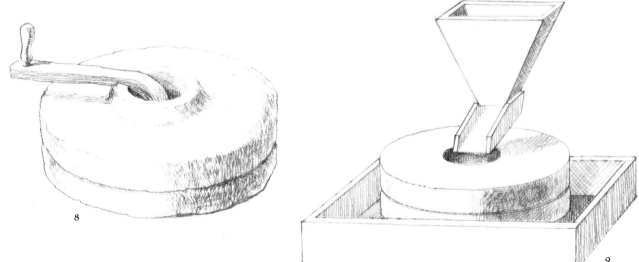

THE DEVELOPMENT OF MILLSTONES

◄ *From the time when prehistoric man discovered that pounding grain with a rock made it taste better, our ancestors gradually developed a method of grinding so effective that it was used for a thousand years. Thanks to archaeologists, we can trace the evolution of the millstone, stage by stage.*

1. Primitive stones for rubbing grain.

2. An early stone pestle and mortar.

3. A saddlestone used in Egypt soon after 3000 B.C., surprisingly similar to other grindstones used in Europe, the Near East, and the southwestern United States.

4. A push mill used in Delos about 500 B.C. has marked improvements: a surface dressed for better grinding, an enlarged hopper for holding grain, and a wooden handle.

5. A Greek mill with a handle that swivels, moving the hopper across a dressed stone.

6. The hourglass mill used by the Greeks fed grain into a cone-shaped hopper to a pair of conical grindstones. The cutaway shows the inner stone, which is stationary.

7. The Delian mill, probably superseded by the more efficient hourglass mill, but significant because it anticipates the French burrstone. Like quartz quarried much later near Paris, volcanic lava provided such a fine grinding surface that it was worth binding fragments into a millstone. Here a lever grinds grain between pieced inner and outer blocks of lava.

8. A fully rotating quern brings continuous circular motion to milling.

9. A "run" or "pair" of traditional millstones. The upper "runner" stone revolves; the lower "bed stone" or "nether stone" is stationary. Above the stones, grain falls from the hopper into the slanted "shoe" which feeds grain into the central "eye" of the runner at a rate that suits the speed of the runner stone and the kind of grain being ground.

The Romans spread these improved rotary hand querns through their empire and began milling on a commercial scale in Italy. They put asses or slaves to work walking treadmills shaped like squirrel cages or huge slabs whose rotation powered the milling machinery. Apuleuis watched slaves working a mill about 170 A.D., wretched men with branded foreheads, half-shaved heads and leg irons: "Their skins were seamed all over with the marks of the lash.... They were covered with flour as athletes with dust." By multiple grinding and sifting, the Romans produced bran, two grades of meal, and one of flour which was sometimes whitened with chalk for patrician customers. Roman loaves were round, flat, and so heavy they sank in water. According to Pliny, the finest flour was "destitute of all flavour," a quality some still prize today.

With plenty of workers, Greeks and Romans were slow to use water power. Greek mills, used in fast-flowing streams, were horizontal wooden paddle wheels set in an upright shaft that turned the upper stone of the small millstones above. These simple wheels probably originated in the Far East, but were so commonly used in northern Europe into the twentieth century that they became known as Norse wheels. Our pioneers housed similar wheels in round wooden casings so that they could run all winter below surface ice. These "tubwheels" inspired nineteenth-century turbines.

We know from Vitruvius' account that the vertical water wheels that grace our old mills were used by Romans before Christ's birth. Usually placed in a steadily flowing stream, the wheel had wooden blades turned by the current. Its movement was transmitted by gears from the horizontal axle to the millstones. Later the Romans developed a wheel which was run by water channelled through a flume and dropped from a height. Roman legions built water wheels through Europe: the Domesday Book of 1085 records 5,624 mills in Britain, most of them watermills. Dams of rock or timber backed up millponds from which sluices and sluice gates controlled water flow. By 1216 these dams and diversions so impeded trade routes that for two centuries English millers had to get Admiralty permission to set them up.

A third type of wheel, the breast wheel, was devised in the sixteenth century. It is turned by the mass of water directed through a lock to strike the blades on one side of the wheel. A variation, the pitchback wheel, discharges its flow from a flume just short of the wheel top. Breast and pitchback wheels turn with the

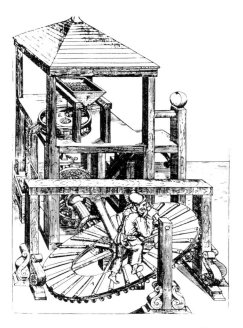

As late as the Renaissance, some mills in southern Europe were driven by manpower. The treadmill shown here is turned by the man who clings to a bar to stay in the same place. Cleat bars round the circumference of the wheel keep him from slipping back.

This type of horizontal mill, used in Provence in the eighteenth century, had a flume to direct a water jet to curved cups.

stream but, like overshot wheels, need a regulated flow of water from a millpond.

Undershot wheels suited coastal tide mills whose gates trapped water at high tide. The miller worked two or three hours while his water supply lasted, then waited for the pond to refill with the day's second tide. Used early along the Massachusetts shore, tide mills were later enlarged so that huge ponds could hold enough water to keep the mill running continuously. In the nineteenth century, Harrington's mill at Canning, Nova Scotia, ground oats with the tides of the Minas Basin, highest in the world.

Undershot wheels also worked well in the middle of a river. Floating mills were either barges with a wheel like a paddle steamer's on either side, driving millstones sheltered by a roof on the central platform, or catamaran-type vessels with a central wheel supported by pontoons. When the Goths destroyed Rome's mills in their attack of 536 A.D., defending commander Belisaurius floated mills in the Tiber that supplied flour even though the Goths tried to jam the wheels with the bodies of their dead. Moored under bridges in mediaeval European cities, floating mills were a contentious hazard to shipping. North America had so many waterfalls that floating mills were scarcely necessary. Paris models sent to New France in 1721 were never put to use in the St. Lawrence River. Floating mills built by Jonathan Devol on the Ohio River and Charles Johnson on Shoal Creek in Texas were regarded as local curiosities.

Gradually millwrights found ways of getting more power from their wheels. They varied width and diameter to suit their water supply, building large narrow wheels for a small stream with a high fall, wide wheels small in circumference for a bigger volume of water with little fall. They replaced paddles of overshot and breast wheels with buckets that held the water through part of the wheel's turn. An overshot wheel carried the weight of the water longest, through a third of its revolution.

Not till the seventeenth century did millers begin to understand that overshot and breast wheels, using gravity, were more efficient than the busily whirling undershot wheel that depended almost solely on the speed of the stream. A century later these inklings were confirmed by John Smeaton whose experiments showed that the undershot wheel was about 30 per cent efficient, the overshot 60 per cent, and the breast wheel halfway between. Combining scholarship and practical experience, Smeaton explored and explained the hydraulics of milling so that dams and sluices could be precisely engineered.

An Armenian floating mill of the type used by Belisaurius to feed Rome under siege.

COMMON TYPES OF WATER WHEELS

1. An undershot wheel, here used in a ▶ tidal mill. At low tide, the miller raises the sluicegate from the millpond behind the mill and the water stored at high tide is gradually released to turn the wheel. At high tide, the sea surges up to the top of the mill's stone foundation and the wheel is immersed almost to the axle.

2. An undershot wheel, driven by water flowing through a sluice from a pond and down a curved channel behind the wheel.

3. A high breast wheel, fed from above the axle.

4. A low breast wheel. Water from the sluice strikes buckets below the axle.

5. An overshot wheel fed by an iron pipe flume.

6. A pitchback wheel. The flume drops its water just short of the top of the wheel.

1

2

3

4

5

6

He also introduced a material that made other improvements possible – cast iron. More resistant than wood to immersion, iron was soon used for wheel hubs, rims, axles, and carefully curved buckets, and later for entire wheels and much of the gearing inside the mill.

In 1824 the French engineer J.V. Poncelet used iron to design an undershot wheel almost as efficient as a good overshot wheel. Sheet metal buckets on a light, rapidly turning wheel were shaped to make double use of water pressure, first as it was fed smoothly into the buckets and again as it dropped out into the tail race, the water running downstream after it passed through the wheel. Three years later, Poncelet's compatriot Benoit Fourneyron would apply this principle of channelling water power in two directions, using both impulse and reaction to turn force into energy, in his invention of the vertical turbine.

HARNESSING THE WIND

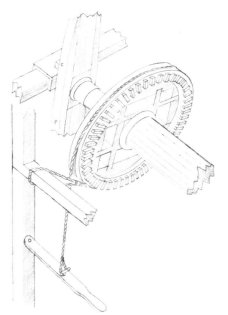

Water is such a familiar power source that its use seems inevitable. Harnessing the wind has an element of magic. Water millers dreaded floods, but their streams flowed along courses they knew, and by building stronger dams, canals, and wheels they learned to bring their power source under control. A windmill, turning to catch power that might come from any direction, was always at the mercy of the weather. The miller had to work when the wind blew, day or night, at a safe velocity of ten to twenty-five miles an hour. In a calm, his sails hung idle. In a storm, the wind became his enemy, ripping off sails, tearing windshafts loose, overturning small mills or pelting them to pieces. If a gale blew up while the mill was running his situation was perilous. A watermill could be stopped by a gear that disengaged the machinery from the water wheel. To stop a windmill, the miller had to use all his strength on a lever that pulled a brake band round the main drive wheel, a wooden wheel up to eleven feet in diameter that could easily catch fire from the friction of the band. A high wind might blow the sails backwards with more force than the brake could hold. Thus tailwinded, the mill machinery raced out of control, heightening the danger of fire from over-heated gears or stones running without grain.

At twenty-eight, singlehanded, George Kimball runs the windmill at Prescott Farm, Middletown, Rhode Island. Built in Warren in 1811, it was moved three times across frozen bays to serve other towns before being brought to Middletown in 1970 by the Newport Restoration Foundation. Now Kimball is back-breeding whitecap flint corn to grind into old-fashioned Rhode

Below left: The brake wheel of a windmill is mounted on the windshaft just inside the mill wall. When the miller wants to stop grinding, he pulls down the lever at bottom left, which tightens a rope that runs in a groove round the brake wheel, bringing it to a stop and, with it, the sails. The brake wheel serves a double purpose: as well as stopping the mill, it corresponds to the face wheel in a watermill, transmitting wind power to the millstones through a series of gears not shown in these drawings.

Below: A more sophisticated braking system lets the miller stop the sails from a lower floor. Here the band round the brake wheel is wooden. When the brake is off, a space of several inches between the two ends of the band allows the wheel to turn freely. When the miller applies the brake, the band is tightened round the wheel by means of a chain running through a system of levers and pulleys down to the grinding floor.

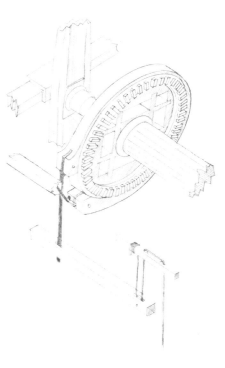

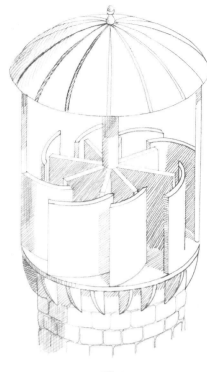

A horizontal windmill designed by Veranzio in the seventeenth century. The fixed outer guide blades and inner vanes that turned with wind coming from any direction anticipated the principle and configuration of some nineteenth-century water turbines.

Island johnnycake meal. He says, "In a windmill you have to have an instinct to know what's going on on the top floor when you're on the bottom floor. Everything is controlled from the first floor, blindly. It all depends on the amount of wind. As the vanes pick up speed you must judge whether they're going too fast, if a gear tooth is out, if the stones are burning. The secret of milling is to keep the stones close enough to grind fine and far enough apart so they don't touch and burn. Your adjustment of the vanes, the shoe that feeds grain to the stones and the stones themselves is all done by sensing, feeling, hearing the rumble and tremble of the whole mill. You never know what to expect. Three years ago I broke my back when I fell off a vane; you have to climb the ash slats like a ladder to set the sails on this mill. It's a challenge every time you grind grain."

Up the road from Prescott Farm, Clifton Boyd remembers how he used to turn the eight-sailed windmill his family has owned for 170 years. "I was running that mill when I was fourteen years old. It was a good deal like sailing a ship. You had to keep one eye on the weather, especially in winter, and we ran it winter and summer. When the wind changed you stopped the mill right away, took in the sails and turned the top to face the wind again. You couldn't run in a thunderstorm because the wind could change too quickly. You had to let the brake on easy. Sometimes if you got caught in a very high wind you could put the brake on and the mill would keep right on going until the gust stopped. Then you would reef in the sails a little more. You could reef our sails standing on the ground but with eight vanes, eight set of sails, it was a long job."

The earliest windmills appeared in flat countries either too dry or too wet to rely on flowing water. Windmills built in Persia in the tenth century were horizontal direct-drive machines much like a horizontal watermill turned upside down. Here, where the wind was constant, it turned vanes set in a round housing on a shaft running down to the millstones below.

The vertical windmills we know were invented in northwestern Europe in the twelfth century. The earliest known picture of one is in the Windmill Psalter of about 1290, once owned by William Morris. Later manuscripts, carvings, and other graphics show them as a familiar part of the landscape.

These mediaeval windmills were post mills, simple frame boxes set on a stout centre post that allowed the whole body, or "buck," to turn so the sails faced the wind. The post was set in a pair of heavy oak cross-trees and braced with four quarterbars.

A post mill.

The buck was supported by a massive oak beam, the "crown-tree," which pivoted on top of the centre post so that it could rotate full circle. The miller climbed into his mill by a ladder hinged to a platform at the back. To turn the mill he raised the ladder foot, put his shoulder to a pole set in the back of the building and pushed the buck round to face the wind. Later mills had a cartwheel on the end of the tail pole, or used horses to turn the buck. The front of the mill carried the sails, mortised to a windshaft which, like the axle of a water wheel, was geared to drive the stones. Eventually millers enclosed the underframe to protect the beams and provide storage space. The whole unlikely structure had to be so light and well-balanced that it could support whirling sails and house one or two pairs of millstones, heavy wooden gears, sacks of grain, and a fairly solid miller. Running fast, post mills surged and pitched like sailboats. Astonishingly, their average working life was about two hundred years.

The fifteenth century brought a dramatic improvement – a stationary mill with a revolving cap. The stone or brick tower mill was topped by a rotating timber cap that carried sails, windshaft, and brake wheel round with the wind. Because only the workings directly connected with wind force had to move, the building was much safer and more stable. The mills of New France were modelled on towers now more easily found in Europe than in Quebec. Tower mills probably originated on the Continent, and the earliest were low enough to let the miller reach sails and tail pole from the ground. In later mills, he hauled on an endless chain that turned the cap. As mills grew taller to catch wind high above trees or other buildings, and to house up to seven floors for various stages of milling (and often the miller's family), they sprouted galleries from which the cap could be turned and the sails adjusted. Some Dutch mills were more than a hundred feet high.

The smock mill, characteristic of England, the Netherlands, and the seaboard states, is essentially a tower mill that suits countries with plenty of woodland and, in some places, a subsoil too soft to support heavy masonry. Smock mills have a wooden frame covered with painted or tarred clapboarding, shingle, or thatch. Timber construction means that the mill must be polygonal rather than round. Most smock mills are eight-sided.

The task of turning a post mill or the cap of a tower or smock mill with tail pole or chain was gradually eliminated after 1745, when Edmund Lee patented a device that faced the mill into the wind automatically. His fantail was a small wheel with wooden

Squat and cone-capped, the restored tower mill at the north entrance of the Bois de Boulogne in Paris is similar in style to those in New France.

A smock mill.

vanes, mounted behind the cap at right angles to the sails. Spun by the wind, it was geared to pull the cap round to the right quarter. When the sails began turning, the fantail stayed motionless in the dead air behind the mill until a change of wind put it back into action. Its gear ratio was so high that it could turn an entire post mill.

As windmills grew bigger, sail design improved. Mediaeval sails hung flat against the body of the mill, mounted on two stocks set at right angles and mortised through the windshaft. Stubby and symmetrical, they carried the same amount of canvas on either side of the stock, laced through or laid on a light wood frame. Next came the "common sail," turned slightly to catch the wind and set with most of the canvas on the trailing side of the stock. Like a seaman the miller reefed his sails to suit the wind force, furling them when he stopped grinding.

Millwrights found that by tilting the windshaft slightly skyward they prevented post mills from being thrown forward by the weight of the sails. They also got more power from the wind, and could slope the mill walls to a broader base. Again, Smeaton explored the aerodynamics of the sail and discovered that by twisting the sail slightly, he could improve its efficiency. He experimented with five, six, and eight sails; his five-sailed windmill at Newcastle, the first in England, was least useful because one damaged sail threw the mill off balance, while mills with an even number of sails could run on four, three, or even two. His use of iron for windshafts made it easier to hold the stocks. Iron gears, usually combined with wood for quiet running, were commonly installed from the late eighteenth century.

The circular fantail was set with slanted blades that responded to the lightest shift of the wind. A shaft from its axle transferred its motion through a series of gears. The biggest gear, fixed to the cap, turned the smaller gear below which meshed with cogs around the base of the cap. The cap revolved smoothly on its track of roller bearings until the sails once more faced the wind and the fantail hung idle in the dead air behind the mill.

THE GRIST MILL: GEARS AND STONES

Although wind and water power could be adapted to dozens of uses, the classic mill is the grist mill. Its wooden gears have the splendid symmetry of ancient machinery aptly suited to function in a tight, oddly shaped space. Forever being mended, replaced, improved, the equipment is endlessly various, with auxiliary gears, belts, and pulleys running sack hoists, conveyors, and sifters from the main power drive.

The axle of a water wheel runs through the mill wall to the parallel face (or pit) wheel. The inner side of this face wheel is studded with wooden pegs that engage with the wallower, the gear mounted at the foot of the vertical main shaft. Early wallowers were either smaller pegged gear wheels or lantern pinions, a ring of staves fitted between two wooden discs. Later wallowers were set with bevelled wooden or iron cogs. The wallower rotates the main shaft to drive an enormous central gear called the great

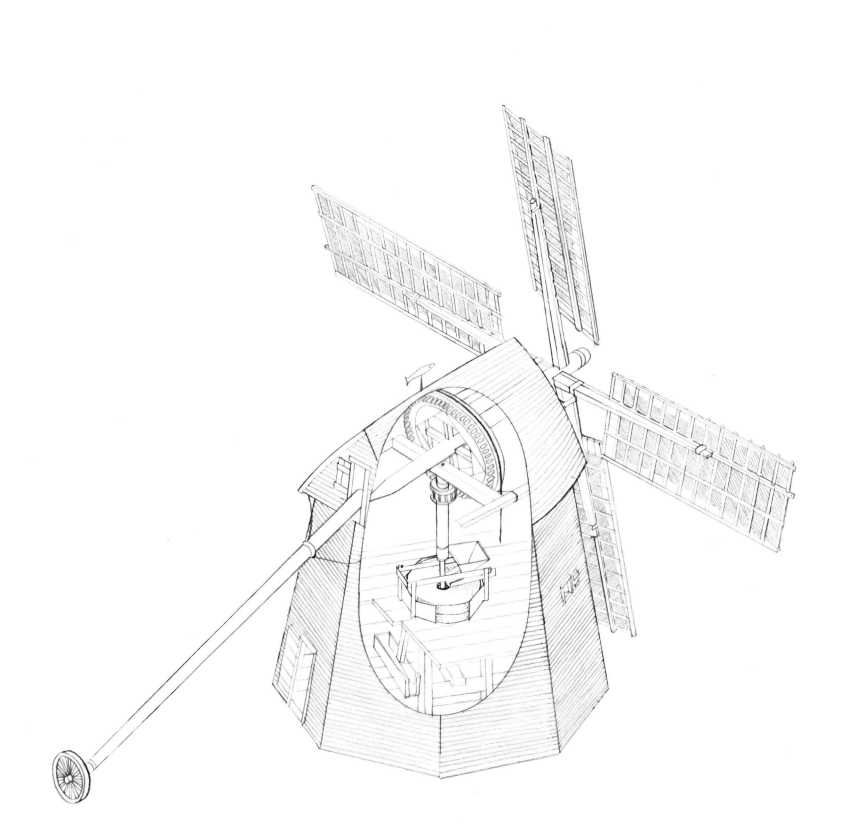

HOW A WINDMILL WORKS

◀ *Though some later North American windmills held up to four storeys under the cap, those built by the pioneers were usually simple affairs in which the brake wheel turned a wallower on a shaft directly rotating a single run of stones below. The miller stopped the mill by pulling on the rope round the brake wheel. To turn the cap round he pushed the tail pole on its small wheel. The boat-shaped cap and fish weathervane were common in the seaports of Cape Cod and Long Island.*

spur wheel; the cogs of the spur wheel drive the millstone spindle, or mesh with one or more satellite gears, each of which drives a pair of stones or supplementary machinery. The spur and other wheels are traditionally made from two layers of oak laid cross-grained for double strength and pinned together with dowels. Cogs carefully chiselled from seasoned hardwood are oiled with tallow. A wheel with an even number of cogs engages with one with an odd number; the extra "hunting cog" prevents uneven wear because the gears mesh differently each revolution.

The drive system of a windmill is much like that of a watermill. Power from the sails is carried by the windshaft to the inner face wheel, here called the brake wheel because the miller uses it to stop the mill. It meshes with a wallower at the top of the main shaft, driving the great spur wheel below.

Each pair or "run" of millstones is housed in a round or octagonal wooden vat. A spindle running up through the centre of the bed stone is topped by a mace-head attached to an iron bridge called the rynd, which is notched into the sides of the centre hole, or "eye," of the runner stone. As water wheel or sails revolve, their motion is transferred through the gear train and spindle to the rynd that turns the runner.

Grain is fed to the stones from the hopper poised above them. The hopper is a wooden box like an inverted, truncated pyramid, supported on a framework called the horse. Through the open bottom of the hopper the grain falls into the shoe, a narrow wooden trough slanted to pour grain into the eye of the runner. The rate of flow is controlled by the angle of the shoe, adjusted by a string attached to its tip, and by the damsel, a forked iron rod straddling the rynd. As it turns with the rynd, it taps the shoe, shaking out enough grain to fill the stones with each revolution. His ear tuned to the clatter of the damsel and the humming of the stones, the miller judges precisely how to set the flow. From the runner eye, grain moves outward through grooves to the circumference of the stones and is swept by a brush or wooden scoop into a hole at one side of the bed stone down a chute to a bin or sack on the next floor of the mill.

The true run of the stones is critical. If they touch in motion, they can be badly damaged or even spark a fire. The distance between them must be set exactly to suit their speed and the type of grain being ground. The bed stone can be wedged up from below to keep it perfectly horizontal. Bearings at the base of the spindle are connected to a screw-nut on the mill floor that the miller adjusts to "tenter" the runner so that it is correctly dis-

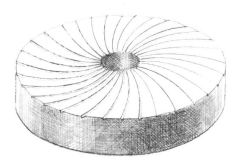

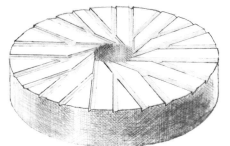

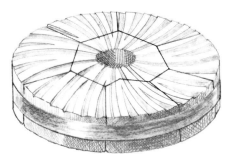

tanced. To balance the runner, millers poured lead into holes in its top, adding more whenever it ran lopsided. Later, adjustable balance weights were set into the stone. Eighteenth-century millwrights devised a lift-tenter that kept the space between the stones constant in proportion to their rate of rotation, and in 1787 Thomas Mead combined this system with a centrifugal pendulum to produce an automatic governor that was soon adapted for use in mills run by Watt's new steam engine.

English stones were solid granite or sandstone but even the best native stones quarried in the Derbyshire Peaks were generally used for oats or barley. For wheat flour, European stones were imported: blue "cullin" stones from Cologne or, finest of all, French burrstones pieced from small bits of quartz, joined with cement, bound with iron hoops and backed with plaster of Paris. Later, composition stones of emery or carborundum proved cheap and durable. Millers tried various speeds and stones ranging up to seven feet in diameter. By the eighteenth century four-foot stones a foot thick, weighing perhaps a ton and turning at 125 to 150 revolutions per minute, were most common.

The working surfaces of runners, and sometimes bed stones, are slightly concave to grind the grain more finely as it works from eye to edge. When the stones are dressed, the runner stone is lifted off and turned over by means of a crane, grappling hooks, and a screw hoist. A wooden bar painted with red lead is passed over the face of each stone to reveal uneven high spots. Then the two stones are dressed in the same direction so that between them they shear the grain like scissors. A complex pattern of flat "lands" and grooved "furrows" provides passage and ventilation to cool the grain. The miller or travelling stone-dresser chipped the furrows in the stone with a mill bill, a sharp double-ended wedge-shaped pick set in a wooden handle. Each man had his favourite pattern, from simple sickle grooves to intricate variations on Roman styles. Usually he began by dividing the surface into ten "harps" with lines drawn at a tangent to the eye of the stone. Within each harp he pecked out furrows about one-and-a-quarter inches wide and half to three-quarters of an inch deep, flat on one side, sloped on the other. Next he "stitched" the lands with shallow parallel cracks closely set like the lines on a file, to skin the bran from the wheat kernel. The whole process kept him steadily at work for two or three days. As his bill struck the stone, bits of steel would fly off and lodge in the skin of the stone-dresser's left hand, and "showing his metal" to a miller was his unwritten diploma. In a busy mill, stones wore

FROM GRAIN TO FLOUR

This early nineteenth-century grist ▶ mill is typical of the period in which millers were replacing hand labour with automatic machinery. The old-fashioned sack hoist (1) lifts incoming loads of grain to the top storey. It passes through a revolving cylindrical screen (2) that whirls chaff and dirt into the sack (3) below the screen. The heavier grain falls through the meshes of the screen to the smutter (4), a machine designed to cleanse the grain of smut, a fungus that could destroy whole crops until smut-resistant strains were developed. Here the grain is also scoured to remove remaining dirt. It drops to the wheat garner (5), a bin that stores wheat so that it can be fed to the hopper (6) at a rate that suits the millstones.

The runner stone (7) is turned by a gear train that transmits power from the water wheel.

Ground meal or flour falls through a chute below the bed stone (8) into a bin (9) from which an elevator (10) carries it back to the top floor to be cooled by the hopper boy (11). Driven by auxiliary gears from above, this rake-like machine with its splayed wooden teeth does the job once done by an apprentice. As it revolves, it gradually sweeps the freshly ground grain from the side toward the centre, cooling it and feeding it down a central chute to the bolter (12). In this slanted cylinder the finest flour passes through closely meshed silk to the sack on the right. As the meshes grow increasingly coarse, middlings, shorts, and bran drop into the second, third, and fourth bags.

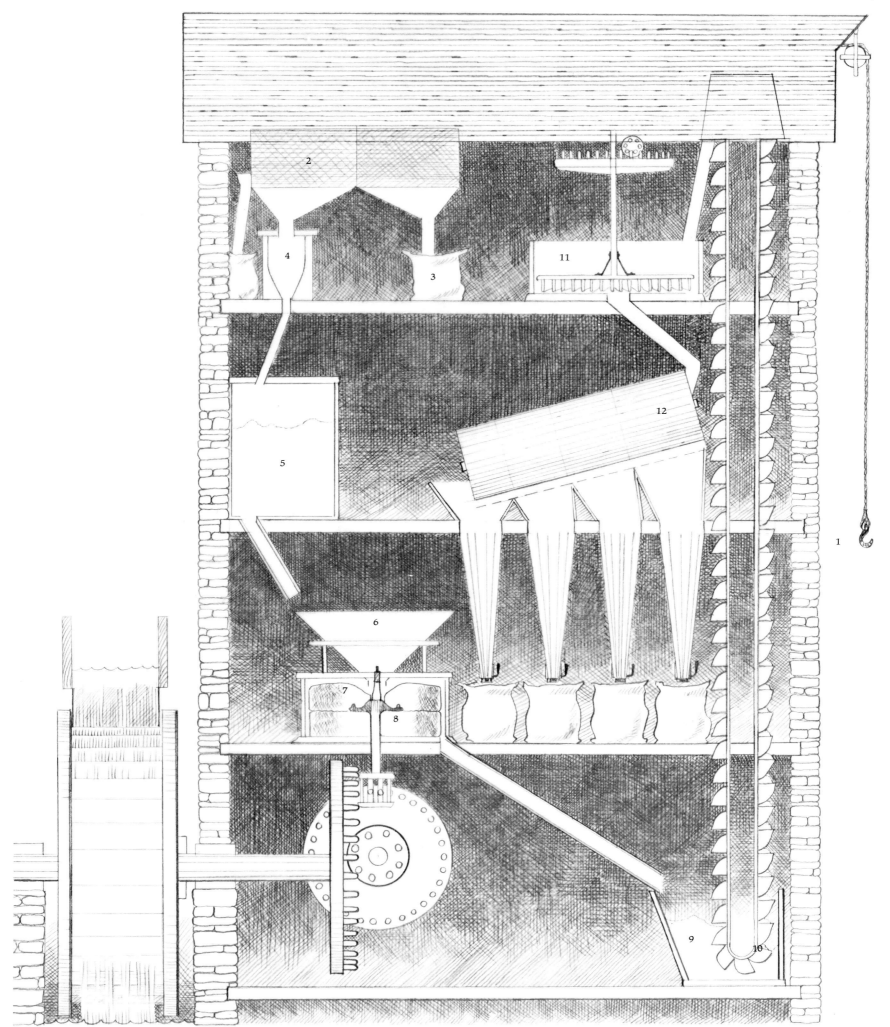

45

A stone dresser's tools rest on the stone.

A detail of four-quarter dress.

down a quarter-inch a year and a pair of well-used French burrs needed dressing every two weeks. Millers and stone-dressers knew that dull stones crushed grain into clammy, lifeless flour.

Archie MacDonald, who owned the Balmoral Grist Mill near Tatamagouche before it was bought by the Nova Scotia Museum is one of the few men who can still dress a stone. Helping his father run the mill from age ten, he was taught by a millwright from West Branch. "William Hingley, he trained me for dressing the stones, the same burrstones we're grinding flour with now. They were brought from France as ballast for timber ships. Your knuckles would be all blood. If the stones need dressing now, I do it: you've got to take them up every year to oil the spindles. I don't know another soul around who knows how to dress a wheatstone."

After grinding, flour was passed through sieves to separate it into degrees of fineness. Early sieves were simply set in boxes which were shaken by hand. Later a crank was used to shake the grist down an inclined trough whose bottom held sieves with meshes increasing in size, so that the finest flour fell into a bin under the first section of the trough and coarser "middlings" sifted into bins toward the lower end. In 1502, Nicholas Boller of Austria connected the sieves with the drive power of the mill, using a series of cloths of different meshes to sift out the various grades of flour. The sieves became slanted revolving cylindrical or polygonal "bolters," and by the end of the eighteenth century bolting cloth was tightly woven not of wool or linen but of silk with 125 threads to the inch. In well-planned mills, grain ran directly from the grindstones to bolting machines underneath.

Grist milling was becoming a fine art, and the miller's social position rose with it. In mediaeval Europe he worked for the lord of the manor, whose peasants were compelled by soke right to have their grain ground at his mill. Soke rights, also held by religious orders, were gradually allowed to lapse from Richard II's reign in England, but remained a grievance in France right up to the revolution. The miller's wage was a fraction of the grain, which he could sell or feed to his family, pigs, or poultry. Like the miller in Chaucer's *Reeve's Tale,* he was often suspected of taking more than his toll. Most millers were probably honest but they were set apart from other tenants by an occupation that called for finer judgement and more mechanical aptitude. From father to son they passed down the craft of dressing stones, rigging repairs, gauging the quality of wheat with their teeth and of meal with left thumb and fingers. In the fifteenth century, European millers began forming guilds. In the sixteenth, English millers were for the first time allowed to become merchants, dealing in grain and meal beyond the amount taken in toll. They made at least twice as much as a labourer, with hope of profits when grain prices rose. By the eighteenth century, some were accused of manipulating the market, buying cheap from farmers and selling dear to bakers. Once they had lived inside the mill; now handsome mill houses reflected their community standing.

OTHER USES OF POWER

Like flour millers, businessmen in other industries were thriving on the proceeds of water and wind power. This was the period, just before the advent of steam, when the potential of the traditional mills was fully realized. They carried out all the processes of cloth manufacture – carding, spinning, weaving, and fulling. They smelted, forged, rolled, and slit metal. And, most important for the pioneers still thrusting west across North America, they sawed lumber.

The earliest sawmills, used in Germany and France from the fourteenth century and through the Continent by the sixteenth, were reciprocating machines, so simple to build and run that they were still common after the circular saw was introduced about

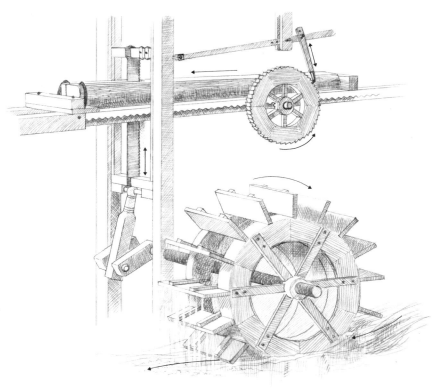

A reciprocating saw driven by an undershot water wheel. The saw blade is tightly held in a framework much like that of a sash window. The axle of the water wheel turns a crank attached to a "pitman," or connecting rod, that moves the sash frame and saw blade up and down with each revolution. The frame slides through grooves in the fixed vertical posts, the saw "gate." The lever harnessed to the top of the frame transfers the up-and-down motion of the frame to the "hand" at right which engages with a notch on the "rag wheel." The inner gear of the rag wheel meshes with the cogs of the log carriage to push the timber, ratchet by ratchet, against the blade as the saw makes its downward cutting stroke.

A paltrok mill, the Netherlands saw-mill extended at the sides to make room for tall timbers.

1820. A vertical blade set in a wooden frame was moved up and down by a crank and connecting rod from a water wheel that also powered a sliding carriage that inched the log against the saw. Slow as hand sawing, the process was later speeded by the gang saw in which the frame held several parallel blades that sliced a whole log into planks in a single run.

From 1592 the Dutch used wind to drive sawmills. Ungainly but efficient, these *paltrokmolen* had wings on either side to house the log carriage. Closed at the front, open at the back, the whole building turned on a circular track to face the wind.

Only in England were sawmills bitterly opposed by hand sawyers. Their threats drove a Dutchman who built a sawmill near London in 1663 to abandon it. As late as 1767 they actually tore one down, but it was an empty gesture since the Thames bank at Lambeth was already lined with post mills for timber yards.

Each type of mill developed machinery to suit its particular purpose. Grist mills needed horizontal stones for the precise grinding of various grains; for materials that had simply to be crushed, vertical stones worked better. In edge runner mills, the millstones revolved vertically round a plate of wood, stone, or

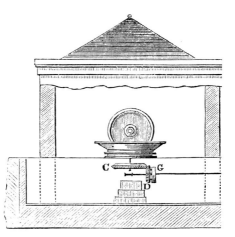

Vertical edge-runner stones, illustrated in Sir William Fairbairn's Treatise on Mills and Millwork.

Its reflection shimmering in the millpond, Balmoral Mills in Colchester County has been restored by the Nova Scotia Museum. In 1873, Alexander McKay paid $12 for the mill site.

iron. Roman slaves had driven stones like these to extract oil from olives. From the seventeenth century the Dutch used them in windmills to crush peanuts. The meal was then packed into coarse bags, placed in a wooden press, and pounded with a ram. The oil ran into a pan and the remaining meal cake served as cattle food. A similar combination of mill and press, both powered by wind or water, produced linseed oil and animal feed from flax seed.

Later, edge runner mills ground cement, flint for china making, and gunpowder. Chart Powder Mills at Faversham in Kent supplied Nelson at Trafalgar and Wellington at Waterloo. (Nelson also shipped a few barrels of flour, not for food but to whiten the wigs of officers wearing dress uniform into battle.) Smaller mills, some with heavy mace-headed pestles instead of pairs of stones, crushed fine powders such as drugs, snuff, mustard, cocoa, and paint pigments.

Windmills were occasionally used for ventilation. Jail fever, bred by overcrowding and foul air, killed countless inmates of Newgate Prison; in 1752, after four judges had died, a fourteen-foot eight-bladed fantail was set over the building to pump air into courtrooms and cells. Even in the nineteenth century, convicts sometimes had to walk treadmills connected to wind vanes that blew the opposite way to double their punishment.

More reasonable, and just as ancient, was the use of mills for lifting water. The ancestor of the Roman water wheel was the noria, a wooden wheel slung with clay pots that scooped river water up to the parched lands of the Middle East. Northern Europe had a different problem – too much water. Here the same principle was applied not to irrigation but to drainage.

The Netherlands are a triumphant example of man's ability to create a country below sea level by using the power of nature to defy nature itself. Before 1000 A.D. Holland was an almost uninhabitable swamp; till 1400 it was ravaged incessantly by the tides and gales of the North Sea. Dikes built in one area would imperil another. Early in the fifteenth century, the Dutch put their windmills to work at drainage, setting post mills above vertical scoop wheels enclosed in brick or timber. Soon they developed the wip mill, a hollow post mill with a small cap holding the brake wheel and wooden lantern-pinion wallower, and a thatched base housing the pit wheel on one side and the miller's family on the other. By 1600 they were building polder mills, powerful smock mills that raised water with either scoop wheels or Archimedes screws.

This engraving from De Re Metallica, ▶ published in 1556 by Georgius Agricola, illustrates a rag-and-chain pump driven by an overshot wheel, the most common method of mine drainage in the sixteenth and seventeenth centuries.

Smock mills began draining the flat fens of England's eastern counties in the late sixteenth century. Until the 1940's some were still at work in the Norfolk Broads where, as in Holland, mediaeval peasants cutting peat for fuel had left lakes and channels that could flood the countryside.

From the late Middle Ages water wheels were used to drain European mineshafts, and more sophisticated systems bred enormous installations. The biggest pumping wheel in the world was the Lady Isabella, built in 1854 to serve base-metal mines at Laxey in the Isle of Man. The great Laxey wheel, seventy-two-and-a-half feet in diameter and weighing a hundred tons, still stands as a monument to Victorian engineering.

Water wheels devised to supply fresh water to growing cities were superseded by complex pumping systems like those at London Bridge and Marley. Here on the Seine in 1682 Louis XIV built a waterworks driven by thirteen water wheels to feed his palace gardens at Versailles. The decline of the French monarchy is suggested by the contrast between the scale of his enterprise and the tiny grist mill in those gardens ordered by Louis XVI as a plaything for Marie Antoinette.

The toy watermill built at Versailles by Louis XVI for Marie Antoinette.

VIEWS OF THE PAST

1. The Lang Mill north of Keene, Ontario, was one of the last in the district to stop grinding. Now it serves as a museum and playground for swimmers and fishermen.

2. The Old Mill Museum beside the falls of Mill Creek in Youngstown Township, Ohio, rests on a stone foundation taller than its three-storey timber top.

3. Mills are usually designed from the inside out, with elevator towers and loading bays built to suit the machinery inside. After the third storey of the Isaac Ludwig Mill in Grand Rapids, Ohio, was destroyed by fire, it was replaced with asymmetrical but useful additions.

4. Though its turbine is idle, this sawmill in Breadalbane, Prince Edward Island, is in better shape than some working sawmills.

5. Watson's Mill at Manotick, Ontario, built west of the Rideau where the river splits to flow around Long Island, is a splendid five-storey grist mill made of limestone from the river bank. Its first owners, Joseph Currier and Moss Dickinson, could afford the workmanship of master mason Thomas Langrell and millwright Owen O'Connor. Dickinson's Ottawa house is now the South African Embassy and Currier's is the prime minister's residence.

6. Gracefully aging into obsolescence, the cider mill in Franklin, Connecticut, still crushes apples into cider by water power.

1

2

3

4

5

6

The application of wind and water power to large and varied industries called for a new class of worker – the millwright, a mechanic who specialized in building and mending mills. Outstanding millwrights like Smeaton, John Rennie, and William Fairbairn were brilliant practical scientists. Even a country millwright was an authority on power, with a working knowledge of arithmetic, carpentry, metal work, drafting, architecture, machine design, and the construction of bridges and canals. Fairbairn described him: "He would reconnoitre and survey the premises on which he was to work, rule and line in hand, and would stand for hours (much to the annoyance of his employers) before he could make up his mind as to what was best to be done. These preliminaries being settled, his decision was final."

Through the experience of centuries, unschooled millers had built up a fund of knowledge of the powers of water and wind. They learned how to control raw energy, to transmit it through gears, to direct it back and forth between vertical and horizontal planes, to change its speed and force to work machinery of all kinds. In windmills they used principles that underlie modern aerodynamics. Smeaton's twist of a windmill sail anticipated the design of an aircraft propellor.

In the eighteenth century, Smeaton and his most gifted contemporaries became civil engineers who joined forces with scientific philosophers to put the millers' practice on a sound basis of mathematics and physics. The French had perfected automation in clockwork toys, conceits for the rich. The English used technology to build factories, bridges, canals, and railways. They combined the powers of water and heat to create a new energy source: steam. The skies of the English Midlands darkened with coal smoke billowing from the mills Blake called satanic.

When our land was first settled, the alliance of scholars and millwrights still lay ahead. No one knows whether many of the marvellous devices in the engravings and sketchbooks of Renaissance Italians like Agostino Ramelli and Leonardo da Vinci were actually put to work in their lifetimes. What *did* work for the pioneers were the makeshift mills of men too busy to speculate on the part they were playing in the development of practical science.

A rust-pocked padlock safe- ▶
guards the Pantigo Mill at East Hampton, New York.

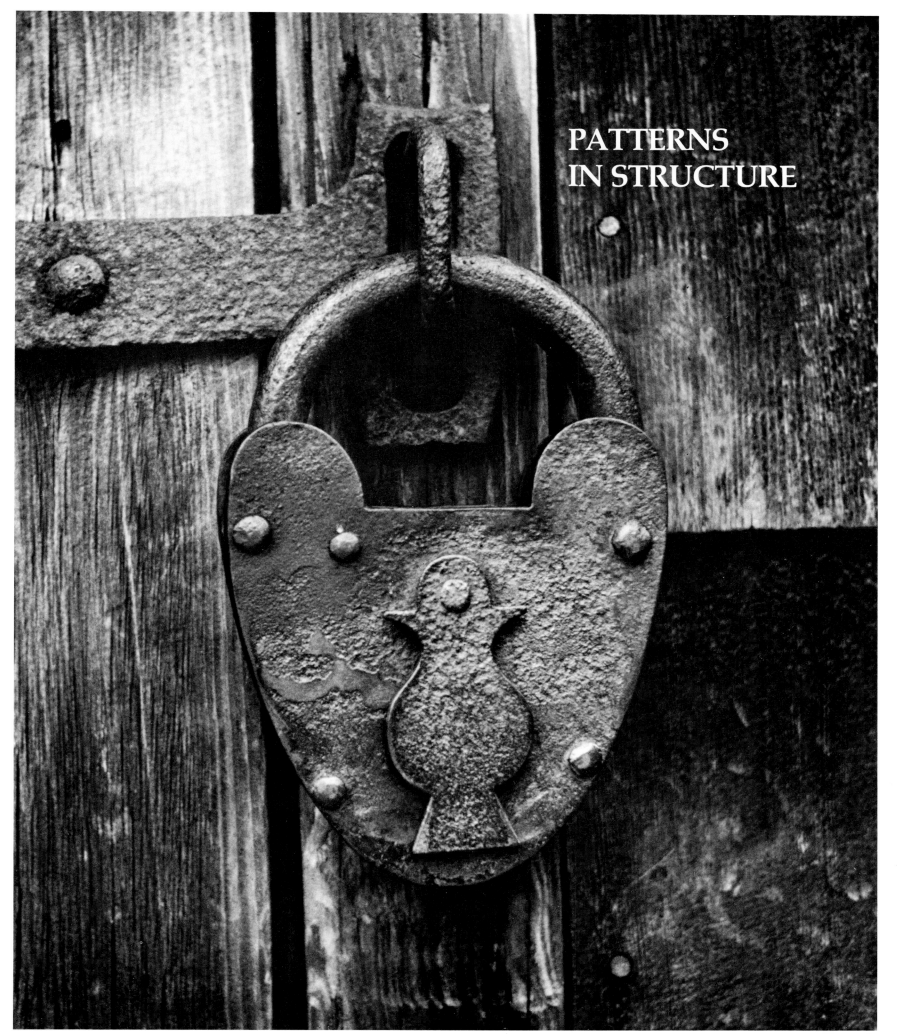

PATTERNS
IN STRUCTURE

1

Doorways

1. This handsome doorway with its neo-classical pediment and pilasters, sidelights and fan transom, once opened into the office of the Pepperell cotton mill at Biddeford, Maine.

2. The doorway of Rosamond Mills, leading woollen manufacturers in Almonte, Ontario. Plain lines and solid masonry reflect the character of the town once called "the Manchester of Canada."

3. Originally a storehouse for a woollen mill, this building in Harrisville, New Hampshire, has a hoist under the extended ridgepole to lift shipments to the attic loading doors.

4. The discreetly recessed doorway of the Crane Paper Company in Dalton, Massachusetts, suits a family-owned mill that has supplied the United States government with specially processed banknote paper for more than a century.

5. The sack hoist over the loft door of the grist mill at Frelighsburg, Quebec, is purely functional. The date above, 1839, is Richard Freligh's proud commemoration of the year he built the mill.

6. At least one of the four generations of Backhouses who ran the mill near Port Rowan must have thought the horseshoe over the door was a kind of amulet. Their luck held; the mill ran continuously from 1798 to 1955, longer than any other in Ontario.

7. The glowing stonework of Allan's grist mill on Grant Creek south of Perth, Ontario, is typical of Lanark County masonry, often the work of Scotsmen who had helped build the Rideau Canal.

8. More than a hundred years old, the flour mill at Ancaster, Ontario, is the fifth on the same site since 1790. In the War of 1812, the mill was used to jail twenty-one American prisoners of war. Eleven were hanged, drawn and quartered; nine were exiled to Tasmania; one escaped and, curiously, moved back to Ancaster.

2

3

4

5

6

7

8

Windows

1. Thick, rough-hewn roof shingles surround a dormer window in the Moulin des Eboulements, built in 1750 and owned for generations by the descendants of Pierre Tremblay who came to New France a century earlier.

2. A window pierces the heavy walls of the seigneurial flour mill built by René Godefroy de Tonnancour in 1721 for his twenty tenant families at Pointe-du-Lac, Quebec.

3. When a mill becomes a museum, everyday tools turn into artifacts to be treasured and displayed. Roblin's Mill at Black Creek Pioneer Village, Toronto, was reconstructed from the five-storey flour mill built in 1842 at Ameliasburgh, Ontario, by Owen Roblin, grandson of a Loyalist from New York State.

4. Only textile mills could afford elegant details such as these Palladian dormers set in the mansard roof of Harmony Mills at Cohoes, New York, most voluptuously Victorian of the great cotton complexes.

5. Though the Rosamonds no longer own this office in Almonte, the ornate stained-glass transom still celebrates the first woollen manufacturer in Ontario.

Textures in Wood and Stone

Mills represent the essence of vernacular design. Architecture seems too grandiose a word to apply to these eminently practical buildings. Clad in local wood, brick, or stone, they blend harmoniously into the landscape.

Only the gigantic nineteenth-century textile factories aspired to architectural presence. Smaller mills seldom squandered time and money on decorative flourishes. Instead they achieve a kind of accidental dignity through their appropriateness to place, period, and purpose. Long use burnishes every surface with a gentle patina. Season after season of sun and snow warps shingles and weathers the texture of fieldstone.

The character of the mills, and the millers, emerges in small details: building methods that strengthen the structure, windows glazed with flawed old glass, idiosyncracies and grace notes added at the miller's whim. Floors are braced with knees hand hewn from tree roots. Roof timbers are deployed with the empirical audacity of mediaeval cathedral vaulting. Lofts hold obsolete machinery preserved simply because it was fashioned by the miller's grandfather. On the porch roof of an abandoned grist mill, a spotted wooden pony prances as merrily as he did when the grindstones went round, long ago.

1. The Old Stone Mill at Delta, Ontario, with its handsome stone-arched windows, is being restored by the Delta Mill Society.

2. At Saint-Antoine-de-Tilly, Quebec, the carding mill is clad in diagonal weatherboarding to shed rain.

3. The Hope Sawmill near Keene, Ontario, is an oddity, faced with stone on the Indian River side and sheathed on the land side with knot-holed, misaligned boards that look like customers' rejects. Built in 1836 by Squire William Lang as a carding mill and family home, it was converted to a sawmill by his son-in-law Richard Hope and worked by Hopes until 1966.

4. Warped by sun and storm, the shingles nailed down by Frank Mansur still protect his mill in Weston, Vermont. Half the houses in town were built with Mansur's lumber before he sold the mill to the Vermont Guild of Oldtime Crafts and Industries in 1936.

1

2

3

Anchors and Beams

1. Like a petrified starfish, this iron support clings to the whitewashed wall of Haines Grist Mill in South Whitehall Township, Pennsylvania.

2. An anchor braces the thick wall of the eighteenth-century Moulin de Tonnancour at Pointe-du-Lac, Quebec.

3. An S-bend support on the Lang flour mill in Otonabee Township, Ontario, which was perhaps the first mill to grind Red Fife, the hardy Galacian wheat introduced to Canada by David Fife.

4. Beams are roughly squared with saw and adze at the Saddle Rock Grist Mill at Great Neck, New York, an early eighteenth-century tidal mill restored by the Nassau County Museum.

5. Wooden dowels pin timbers at the restored seventeenth-century ironworks at Saugus, Massachusetts.

6. Roof framing of the flour mill at Lang, Ontario, was rebuilt after the mill was gutted by fire in 1896.

7. Sunlight illuminates the timbers of the loft in the Loyalist mill at Ball's Falls, Ontario.

8. Intricate cross-bracing supports the roof of the banal mill of Seigneur Godefroy de Tonnancour at Pointe-du-Lac, Quebec.

Overleaf: Like most surviving grist mills, the mill at Acton, Ontario, now produces not flour but animal food stacked in bags on the porch. The roof that protects them has been periodically shored up with hewn beams and braced planks.

WE SELL MASTER FEEDS

3

SETTLEMENTS
ON THE SEACOAST

Helen Fox

64

3

MILLS OF THE PIONEERS

The east coast was our first frontier, and each outpost was isolated in wilderness. Most were scarcely aware of the other settlements that shared their hardships: hunger, cold, building shelter, clearing land, working out rules for safe and harmonious habitation. Sowing the first crop, that momentous step that anchored our ancestors to the New World, was simply another chore. But everywhere, the opening of the first mill was an occasion for celebration. It symbolized independence from the homeland and marked an end to hand grinding and hewing.

Such was the significance of the first mill in North America, a grist watermill built in 1607 by Jean de Biencourt, Baron de Poutrincourt, on the Rivière de L'Equille at Port Royal in Acadia. Poutrincourt's lawyer, Marc Lescarbot, who accompanied him, wrote, "And considering how much toil the hand-mill gave us, he ordered a water-mill to be made, much to the admiration of the savages." Poutrincourt opened it with a ceremony to seal his friendship with the Micmac chief, Membertou.

It was a day of triumph. The colonists of Sieur du Gua de Monts had survived the deathly winter of 1604 on an island in the Ste Croix River and moved to the hospitable Annapolis Basin where De Monts had given Poutrincourt title to Port Royal. Poutrincourt followed Champlain's example by establishing an Order of Good Cheer which kept his table "joyous and well provided" with fish and game. Next spring the mill yielded not only bread but herring, since the sea at high tide carried them up to the mill where they were trapped by the millers.

Soon the cheer faded. The king cancelled De Monts' charter and Poutrincourt sailed for France, leaving ten barrels of meal and the standing wheat for Membertou. "And in this business," wrote Lescarbot, "our water-mill came in extremely useful, for otherwise there would have been no way of preparing sufficient flour for the voyage. But as it was, we had superfluity thereof, which we gave to the savages in token of remembrance." Good sentiment, good sense: when Poutrincourt brought back a new company of colonists in 1610, he found the Micmacs had guarded the settlement.

◀ *The Wolf Pen Mill in Jefferson County near Louisville, Kentucky, has brick-and-stone foundations two feet thick, laid without mortar. Driven by a twenty-six-foot wooden wheel, the mill is still equipped to grind corn.*

A mill of the late sixteenth century, based on a design from Diderot's Encyclopaedia, *has been built on the site of Poutrincourt's mill to house the Lequille Hydro-Electric Station.*

Acadia was a perpetual battleground between French and English. In 1621 James I of Britain claimed and renamed it Nova Scotia. In 1632 it was back in French hands when Charles d'Aulnay seized control, made himself governor and, in the intervals of his ruthless rivalry with Charles de la Tour, tried to settle Port Royal by draining marshlands and building dikes and mills.

Meanwhile, Champlain had thrust inland to found Quebec in 1608. From the first, France's colonial hopes in the New World were torn apart by the lure of the fur trade. From 1628 to 1663 the government of New France was controlled by the Company of One Hundred Associates who encouraged profit at the expense of settlement. The seigneurial system served a real purpose where tenants needed protection from Indians. Champlain had made the critical mistake of choosing the Hurons as his allies and the Iroquois revenged themselves by incessant raids. But few of the noblemen granted land on the St. Lawrence River ever crossed the Atlantic and their vast tracts lay idle. The only seigneurial privilege observed in New France was that of the banal mill, which bound habitants to have their grain ground by the seigneur. Since the toll of one-fourteenth was scarcely enough to pay a miller's wage, even resident seigneurs put up poor mills or none at all. The farmers, who ate two pounds of bread a day, built their own mills at Lévis and Cape Diamond in the 1650's.

Le Moulin du Côteau at Ville Marie was the second windmill built by Maisonneuve, founder of the French settlement that later became Montreal, and governor of Montreal Island. As its name implies, the tower mill was built on the hillside below the crest of Mount Royal. These Montreal mills served the double purpose of grinding grain and providing refuge from Iroquois attacks. Downriver in Huron territory, the town of Quebec was relatively safe.

More industrious and skilled in agriculture were the religious orders with their own banal mills. The Jesuits, granted land at Notre-Dame-des-Anges near Quebec in 1626, brought a windmill from France. By 1650, both Jesuits and Ursulines had sawmills. The orders exercised rights like the seigneurs, and worked harder at colonization. Their tithe, collected only on grain, was never enough to make them self-sufficient. Like the secular landowners, they depended on funds from France.

In 1648, six years after Ville Marie was founded, Maisonneuve built Montreal's first windmill, the *moulin du fort* at Pointe Callières. Ten years later it was replaced by the *moulin du Côteau*. Both mills were stout stone towers, loopholed for defence against Iroquois attack. But more than fear of Indians the quest for furs so retarded colonization that New France in 1666 had only 3,215 settlers – nine of them millers.

Like those of France, England's first aims in settling America were military and commercial. Colonies in Massachusetts were seen as a bulkhead against French encroachment from the north, in Virginia against Spanish from the south. Sir Walter Raleigh, after two vain attempts to settle Roanoke, was still planning expeditions from his cell in the Tower of London when the Virginia Company landed three small shiploads of colonists at Jamestown in 1607. Thanks to the leadership of John Smith, they survived quarrels, Indian attacks, and "the starving time" of 1610 when they ate dogs, mice, snakes, and at least one human corpse. In 1608 Smith brought in Europeans to build sawmills, which his English settlers had never seen, and Richmond, founded in 1611, soon had "three streets of well-framed houses."

Raleigh had charted Virginia's course when he discovered tobacco. From 1614, when John Rolfe planted the finer South American strain, tobacco seized the economy. Grown even in the streets of Jamestown, it became almost the only crop and unit of monetary exchange. It ravaged the soil, but its export ensured the wealth, power, and culture of the colony. Virginians depended mainly on English provisions until 1621, when the first corn mill in the United States was built by Governor Sir George Yeardley on his plantation "Flowerdew Hundred" on the James River.

In time tobacco engrossed Maryland, too. Lord Baltimore's proprietary colony, planned as a refuge for persecuted Catholics as well as a merchant venture, had been well governed from its start, and the first crops were wheat, corn, and other food. But in 1639, when the assembly authorized taxes to pay for a water-powered corn mill, the cost was set at 20,000 pounds of tobacco.

The *Mayflower* brought North America a wholly new kind of settler. The Pilgrims had no royal charter, no backers hoping for profit, no title to the land. They came to Plymouth in 1620 with not much but their bare hands and two ideas that would pervade New England: Puritanism and free enterprise. Ironically, the principles of diligence and thrift underlying their brave attempt to form an orderly society gradually bred a competitive spirit. The Mayflower Compact, a government based on general agreement, never really worked. Communal landholding broke down after drought destroyed their crops in 1623. True to principle, the Pilgrims doled out their last pint of corn, five kernels per person, but Governor William Bradford then assigned each family an acre of land. He wrote, "They begane now highly to prise corne more pretious than silver ... for money they had none, and if they had, corne was preferred before it."

The Pilgrims were followed ten years later by a colony with similar hopes of religious freedom but very different support. The Massachusetts Bay Company originally had a charter from Charles I; but in 1629, when the king dissolved Parliament and intensified his harassment of non-conformists, the company sold its rights to a group of prosperous Puritans who elected John Winthrop governor. Though these stockholders planned to enforce their ideals by keeping control in their own hands, their colonists pressed for representative government and won it in 1634.

The post mill built about 1720 in Williamsburg, Virginia, for William Robertson, clerk of the Governor's Council, was reconstructed near the Governor's Palace in 1957. Typical of early eighteenth century windmills in the Tidewater region, it turns on a centre post hewn from a white oak three hundred years old. Its canvas sails are set by a miller and his apprentice.

Helen Fox

SAUGUS:
Iron, Wood, and Water

The first successful ironworks in North America were planned and supervised by ironmaster Richard Leader for a company headed by John Winthrop, Jr. Leader chose a site with the three essentials for manufacturing iron. At Saugus, Massachusetts, he found plenty of bog iron, timber for fuel, and the Saugus River to power water wheels, ship in supplies, and take products to markets and ports.

Miners dug ore from shallow deposits in marshland, and colliers turned wood to charcoal in slow-burning kilns. Ore and charcoal were dumped through a hole in the top of a huge square stone furnace along with flux (broken rock holding minerals that separated the ore into metal and slag). Three days in a hot fire melted the iron to liquid that collected in a crucible at the bottom of the furnace. The lighter slag rose to the top where it could be skimmed off. Twice a day, workers drew off the molten metal. Poured into moulds, it made cast-iron pots and other housewares; poured into furrows in sand, it made bars to be sold as pig iron or taken to the forge for further refining.

In the forge, sweating men toiled at the job of turning brittle cast iron into stronger, purer wrought iron. Four water wheels drove giant bellows that blasted air to three fires, and a thundering trip hammer that pounded the metal on an anvil. Again and again the metal was heated and hammered until it was fit to make tools or machinery.

Some wrought iron was sent from the forge to the rolling and slitting mill where the red-hot metal was rolled into sheets. These "flats" were sold or reheated and passed through a set of cutting disks that sliced them into "rods" to make nails. Rollers and cutters were driven by a pair of overshot water wheels.

1. The arched entrance to this Saugus furnace leads to the crucible, the basin set in the centre to collect molten iron. The length of the tools at right shows that the ironworkers kept a safe distance when they poked charcoal, skimmed off slag, and ladled out the molten metal.

2. The forge, with the rolling and slitting mill to the right.

3. A pitchback wheel, one of four wheels that drove the bellows and great hammer in the forge.

1

2

1

Working the Iron

The village that grew round the Saugus Iron Works was North America's first company town. Named Hammersmith for the London district where many of the skilled ironworkers had learned their trade, it belonged to Winthrop and the English shareholders whom he had persuaded to invest capital. The company built a handsome house for the ironmaster's home and office, one- or two-roomed clapboard cottages for workers and their families, and a store where they could charge their purchases against their wages.

At the ironworks, artisans recruited from England ran the furnace, forge, and rolling mill. Colonists were hired for rough manual work. Sixty-two young Scotsmen captured by Cromwell were bought by a London shareholder and sent to Hammersmith as indentured servants.

Ironmaking was heavy, dirty, and dangerous. Workers were seared by molten metal, ripped by slag, deafened by the thudding hammer, or crippled from lifting iron bars. Small wonder that some were called into court on charges of swearing, getting drunk or breaking the Sabbath. But many settled down and married into Puritan families.

In the end, the ironworks benefited not the English shareholders but the workers and the colonies. Though the equipment and methods used at Saugus were as up-to-date as any in Europe, poor management and high production costs kept the enterprise from yielding a profit, and it closed down about 1670. Meanwhile workers and their sons moved to other New England ironworks where their skills were highly marketable. Bettering their own fortunes, they helped to establish the iron industry in their adopted country.

2

3

1. Inside the forge, the bellows that feed the hearthfire with its stone chimney are opened and shut by means of a cam attached to the axle of the water wheel.

2. A bar of pig iron protrudes between the forge's 500-pound hammerhead and the anvil below. The hammer is mounted on an oak shaft that swings up and down with the movement of cams driven by a pitchback wheel.

3. Between hammer and anvil the hot metal is pounded into malleable wrought iron.

4. Gears in the rolling and slitting mill drive machinery that flattens wrought iron into sheets, many of which were used to mend or expand the ironworks. Some flat sheets were passed through cutting discs that slit them into narrow rods that could be pounded into nails.

4

SOWING THE SEEDS OF INDUSTRY

Native corn was the first staple diet along the seaboard. Settlers learned from Indians how to hill up their plantings, fertilize them with rotting fish, and grind the dried corn with log pestles and mortars. Townsmen soon recognized the benefits of mills and granted their builders free land, water rights, freedom from taxes, and often a district monopoly.

The Massachusetts Bay settlers sowed English wheat and rye as soon as they landed, and built their first windmill at Watertown in 1631. Their crops were meagre, since the soil was inhospitable to imported grain and they seldom practised rotation. The Watertown mill was being used to grind wheat from Virginia when it was moved to Copp's Hill north of Boston in 1632. Ten years later it was shattered by lightning; the miller lay unconscious twenty-four hours.

Soon the brooks and tides of New England attracted watermills. Dorchester had at least one by 1634 when, "Mr. Israel Stoughton hath liberty granted him to build a myll, a ware, and a bridge over Neponsett Ryver, and to sell the alewives he takes there at 5s. the thousand." John Jenney, who built a mill "for grinding and beating of corn upon the brook of Plymouth" in 1636, was reprimanded two years later for not grinding well and seasonably. In 1639 a town meeting at Dedham ordered the first canal in New England, a three-mile ditch called Mother Brook which drew off a third of the water of Charles River to drive mills. By the next year at least five other towns had mills.

The colonies at Plymouth and Massachusetts Bay had been carved out of a tract from Newfoundland to Philadelphia granted in 1620 to the New England Company, which gave colonizing rights to Sir Ferdinando Gorges and John Mason in the future states of Maine and New Hampshire. The first New England sawmills began attacking the dense Maine forest at Berwick in 1631, and here, in 1650, the first American gang sawmill was built. Mason brought Danish workmen to build New Hampshire's first sawmill at Portsmouth in 1634, but it must have been a poor one. Residents later petitioned the General Court at Boston for a sawmill because "there is none in this Towne, but only one which was never perfected nor like to bee."

The most conspicuous gap in England's seaboard supremacy was New Netherlands, firmly couched in the Hudson-Delaware region from 1624. Like New France, this colony of the Dutch West India Company was a furtrading stronghold. Except for Kiliaen van Rensselaer of Fort Orange, now Albany, the Dutch patroons were no more concerned with settling their lands than

While the Dutch held New Netherlands, windmills enlivened the Manhattan skyline.

were the seigneurs of Quebec. In 1631 van Rensselaer brought in a master millwright, two sawyers, and two small millstones for a corn mill on his estate. Five years later Barent Pieterse Coeymans arrived. Famed as "The Miller," he took charge of van Rensselaer's watermill, built other corn and sawmills, and in 1647 quit the patroon's service and rented and ran his own mills. In 1673 he bought land south of Albany where he founded the town of Coeymans and a dynasty of milling descendants who spread through New York and New Jersey.

At New Amsterdam the Manhattan Island skyline was dominated not by skyscrapers but by post mills, smock mills, and *paltrokmolen*. New Sweden, a tiny Delaware River settlement later annexed by the Dutch, built a grist windmill Governor John Printz called "good for nothing" and replaced in 1643 with a Norse-type watermill.

Dutch rule in the New World lasted only forty years. By 1639 they had lost the fertile and strategic Connecticut River valley to Bostonians led by Governor John Winthrop's son. In 1664 the polyglot population of New Amsterdam surrendered to four English ships in return for the right to keep their customs, trade, and liberty of conscience.

Religious quarrels and the pull of fresh land were already drawing settlers west from Massachusetts. Rhode Island, whose future lay in shipping and manufacturing, was founded on faith. In the Puritan colonies non-conformity bred its own factions, and Massachusetts began expelling dissenters. Banished from Boston, the inflammatory Roger Williams in 1636 established Providence

on the principles of religious freedom and separation of church and state. With tobacco, livestock, and lumber as its first staples, Rhode Island had water-powered sawmills by 1639 but no corn mill till 1646.

Iron was needed everywhere for tools and housewares and for pig iron to export for scarce English cash. Virginia, Massachusetts, and Connecticut had rich deposits of bog iron. Mixed with flux of limestone, feldspar, or oyster shells to draw off impurities into slag, it was smelted in charcoal furnaces, square stone towers whose fires were nourished by bellows powered by water wheels.

The first ironworks built at Falling Creek in Virginia in 1619 was a doomed venture, destroyed three years later by Indians who massacred the villagers. Massachusetts was more fortunate. In 1644 John Winthrop, Jr., got a charter and capital from London for "The Companie of Undertakers for the Iron Works" and brought out master mechanic Joseph Jenks and other English ironworkers to Saugus. Their ten-foot dam across the Saugus River flooded about a thousand acres, raising the water level more than six feet to provide a thirty-foot head of water for a thirty-foot water wheel. By 1646 the furnace was smelting, and pig iron, tools, and utensils were cast from Jenks' models. The Massachusetts government reported in 1648, "The furnace runs eight tons a week, and their bar iron is as good as Spanish." Three years later Scots prisoners captured by Cromwell at the Battle of Dunbar were sent to work as colliers. Though never very profitable, the Saugus ironworks operated until 1670, and the craftsmen imported to run it spread through the colonies to establish the forerunners of the iron and steel industry of today. Winthrop moved to Connecticut in 1646 and built the first ironworks in that state, where eleven years later he became governor.

At Saugus in 1652, Jenks cast the dies for the first colonial coins: shillings, sixpences, and threepences stamped with a pine tree "as an apt symbol of the progressive vigor" of Massachusetts. Shortage of cash plagued the colonies: beaver pelts became legal tender in New York, tobacco in Virginia, Maryland and North Carolina, wheat in Pennsylvania, and rice in South Carolina and Georgia. Eventually they all issued paper money, which depreciated. Britain provided no North American currency and her flow of supplies kept the colonies in her debt. Furs, fish, and lumber exported to Spain, Portugal, and the West Indies brought in Spanish and Portuguese coins, and the Pine-Tree Shillings were designed to keep these in the states. Melted and recoined with less silver than British currency, they were meant for trade only in Massachusetts but circulated even in England.

◄ *The vertical shaft of the Norse mill drives the stones directly above the wheel. In this building a flume directs water to one side of the wheel to keep the blades turning steadily. The lever set in the wall to the right of the grindstones controls the sluicegate to the flume. In this drawing, the end wall has been cut away to show the inner workings.*

TIES WITH
THE OLD WORLD

Great Britain saw her colonies as a source of raw materials and a market for her own manufactures. Even under Cromwell, who considered New England "poor, cold and useless," the home government discouraged colonial textile-making which threatened England's profitable industry. But few settlers could afford imported clothes, and most made their own of wool or linen, using hand cards, wheels, and looms. From the 1640's New England assemblies offered incentives for raising sheep, growing flax and hemp, and the production of cloth.

The heaviest part of the textile process was fulling – finishing the woven cloth by soaking, wringing, and stretching it like a thick rope round the top of a long table where workers pulled, twisted, and turned it for hours. Their labour was eased by the fulling mill in which two heavy oak mallets powered by a water wheel pounded home-woven wool in a bin holding water and soft soap, fuller's earth, or animal urine that scoured, softened, and shrank the cloth to a thick, even texture. The first fulling mill in North America was built in 1643 by John Pearson in Rowley, Massachusetts, founded by fifty-four families led by the Reverend Ezekiel Rogers. A non-conformist turned out of his Anglican parish in Yorkshire, he laid out a town where land was communally owned for nearly five years.

Most of his original settlers were cloth workers. One was Maximilian Jewett, whose descendants moved to New Brunswick in 1804 to found the village of Jewett's Mills which vanished under the Mactaquac Dam in 1967. Through a curious turn of history, the family returned to the woollen business in 1859, adding a carding mill to their trade in grist and lumber to occupy Nathaniel Jewett who had lost an arm in the sawmill.

There's a certain irony in the fact that the Jewel Mill on the site of Pearson's fulling mill now uses the same water power to polish semi-precious stones. The search for the gems, gold, spices, and silks of the Orient drove the first explorers west across the wild Atlantic in their bravely painted cockleshells. What they found – fish, furs, timber, and limitless land – was more mundane, yet in the end more miraculous than any eastern treasure. Marc Lescarbot saw early that, "Our farming is the first mine for which we must search ... for whoso has corn need have naught to do with treasure." Within sixty years, settlers discovered their most valuable asset was plain hard work.

Society in the North American colonies was always fluid but far from classless. France's attempt to maintain the feudal system in Quebec was constantly eroded by the exploits of the

◀ *The feed mill owned by Bill Furs of Marchmount, Ontario, is one of the last to run on a turbine. The first mill on this site near Orillia ground flour for the Indians.*

coureurs-de-bois, the young men who deserted the land for the fur trade. Farming and milling were neglected unless an exceptional leader spurred seigneurs and habitants into action. The banal mills, their history linked with seigneuries and parishes, were venerated only as they crumbled.

The English settlers, on the other hand, bettered themselves by exercising the skills each community needed most – planting tobacco, shipping lumber, trading, and manufacturing. Wherever a mill promised a profit, someone built one; with success, he enlarged his business and his local substance.

Joseph Jenks was one of the immigrants who seized the opportunity to establish his independence. He was forty when he came from Hammersmith to Saugus. Within four years he was petitioning the court "for liberty to make experience of his abillityes and Inventions for the making of engines for mills to goe with water ... and mills for the making of sithes ... with a new Invented sawe-mill." His patent was withheld nine years on the grounds that such manufactures were too valuable to be monopolized. But he persisted, and finally persuaded the ironworks to sell him the right to build his own forge. Here in 1655 he began making scythes "for the more speedy cutting of grasse." His design was so radically different from the old English type that it has been used unchanged ever since. A hundred years later, the commissioner of patents declared that Jenks' improvement was, in its day, a greater mechanical advance than the mowing machine.

Jenks' son Joseph inherited his spirit. When the Restoration of Charles II renewed anti-Puritan persecution in England, young Joseph was hauled before the Massachusetts Court to answer for his "treasonable utterance" that "if he hade the King heir, he wold cutte off his head and make a football of it." Not surprisingly, he eventually moved west to found a new town, Pawtucket. In 1671, he bought sixty acres near Pawtucket Falls and set up a sawmill and the first forge in Rhode Island. His eldest son Joseph, a strapping fellow seven feet two inches tall, was governor of Rhode Island from 1727 to 1731.

The Jenks were just one among many ambitious families who recognized the demand for mill products of all kinds and rose to eminence by providing them. Energy like theirs pervaded colonies now certain that they would endure.

The old mill stream is still a ▶ swimming hole beside the pre-Revolutionary grist mill at Stony Brook, New York.

THE GRIST MILL

The Essential Providers: North American Grist Mills

1

2

3

1. In Clinton, New Jersey, the fourteen-foot wheel of the Old Red Mill runs on water from a dam that stretches two hundred feet across the Raritan River.

2. The mill at Les Eboulements, Quebec, with steeply pitched roof and low dormers echoing Norman style, is now preserved by Heritage Quebec.

3. The Indian Mill near Upper Sandusky, Ohio, was built in 1861 near the site of a grist mill and sawmill provided forty years earlier by the United States government for the Wyandot Indians, who had supported the States in the War of 1812. They were the only tribe given government mills, probably because an agent on their reserve insisted the mills were necessary.

4. Haines grist mill on Cedar Creek in Pennsylvania was built before the War of Independence and extensively reconstructed after a fire in 1905. Windows above the gambrel roof light the fifth-storey attic.

5. The grist mill in the restored ironworking village of Batsto, New Jersey, has the characteristic projecting bay for lifting grain sacks from ground level to loft.

6. The Cornell Mill at Stanbridge East, Quebec, was built by Zebulon Cornell in 1830 and owned for more than a century by four generations of Cornells. It is now the Missisquoi County Museum.

7. Late afternoon sun gilds the fields and red clay roads of North Tryon in Prince Edward Island. The grist mill on the left and sawmill on the right have been owned by the Ives family for a century. When Charles Ives installed the first flour rollers on the Island in 1902, farmers brought wheat from beyond Charlottetown. The sawmill was owned by his brother George, whose son Everett took over both mills in 1933. Now the sawmill is idle, and George's daughter and son-in-law, Frances and Charles Roberts, grind animal feed in the old grist mill.

4

5

6

7

1. Until the dam burst in the spring of 1965, the Pike River in the Eastern Townships of Quebec drove a vertical wheel and later turbines for the mill at Frelighsburg. Here Joseph Gagnon ground buckwheat and animal feed for nearly fifty years. Jean-Marie Demers bought the mill in 1967 and has been restoring it ever since.

2. Though twice gutted by fire, the flour mill at the mouth of the Elora Gorge north of Guelph, Ontario, worked into the 1970's.

3. The Delta grist mill was once owned by John Beverley Robinson, later attorney-general and chief justice of Upper Canada. Robert Gourlay, the Scots gadfly who was expelled from the province for criticizing its Establishment, nevertheless described the mill as "unquestionably the best building of its kind in Upper Canada."

4. Balmoral Grist Mill near Tatmagouche, Nova Scotia, was built in the 1870's by Alexander McKay, who harnessed the power of Matheson's Brook to drive a Leffel turbine.

5. Once a grist mill, Brant's Mill now pumps water for Rolling Rock Estate in Ligonier, Pennsylvania, named for the fort built by the British field-marshal Sir John Ligonier.

6. A corner of the century-old Morningstar Mill is gently subsiding into the spectacular DeCew Falls near St. Catharines, Ontario. Here Wilson Morningstar installed a turbine, five sets of flour rollers, and a thrifty bolting system that saved expensive silk by using oscillating semi-circular sieves instead of fully rotating bolters.

1

2   3

A Heritage Preserved

Whether they grind wheat flour, oatmeal or corn, grist mills have attracted most of the legends of milling. Perhaps because they satisfy our nostalgia for the past, more have been preserved than any other type of mill. Few restored sawmills or textile mills exist outside museums and pioneer villages. Old mills used as houses, shops, craft centres, restaurants, or local museums are almost all grist mills.

Most interesting are those that still carry on their original craft. A surprising number of grist mills are in working order, demonstrating one of the most important aspects of pioneer life, producing animal feed or making flour for customers who consider bread baked from stone-ground flour healthier and more appetizing than mass-produced enriched bread.

Every run of grindstones has its own voice: some sing softly and steadily, others mark each revolution with a low thud like the sound of a distant train.

1. The Wayside Inn grist mill in Sudbury, Massachusetts, was built in 1929 by Henry Ford as part of his grand design for dotting the countryside with village industries. Restored after a fire in 1955, Ford's notion of a picturesque mill is now a National Historic Site.

2. Fowles Mill, built at Hastings, Ontario, in 1850, now sits high above its drained millpond, exposing the rack that screened flotsam from the mill race.

3. A reconstruction of the grist mill built in 1770 by George Washington on his estate at Mount Vernon. A sixteen-foot pitchback wheel on Dogue Run drove French burrstones to grind wheat flour for sale and cullin stones to do custom grinding of wheat or cornmeal for local farmers. Riding to his mill in 1799, Washington caught a chill that was thought to have hastened his death.

4. On the Rocky Saugeen River north of Durham, Ontario, Alex Ferguson in 1851 built his grist mill. Used as a chopping mill for animal feed until 1955, it is now owned by W. H. Lind, a Toronto lawyer who has rebuilt the dam and preserved the mill artifacts.

1 2

1. The Old Mill on Stoney Brook in Brewster, Massachusetts, was built in 1893 to replace Thomas Prence's seventeenth-century mill, burned in 1871. In 1940 the town of Brewster bought the mill and replaced the turbine installed in the 1880's with an overshot wheel.

2. At the Wight Grist Mill, reconstructed on its original site in Old Sturbridge Village, Massachusetts, grain from the hopper pours into the tapered "shoe" below, which feeds it into the eye of the runner stone. The stones are enclosed in a round wooden vat.

3. At Roblin's Mill, a swing hoist and screw lift fitting into holes on either side of the runner raise it from the nether stone and turn it over for dressing.

3

1

2

3

4

1. The "hopper boy" invented by Oliver Evans cools and collects ground meal to be swept down to the bolter at Roblin's Mill.

2. At the Lang Mill near Keene, Ontario, a vertical bolter sifts flour to remove coarse particles for regrinding, and bran for separate processing.

3. Behind the Lang Mill, the conical cyclone duster collects the flour dust from the mill.

4. In Warren Leard's flour mill at Coleman, Prince Edward Island, brightly striped chutes deliver flour sifted by a bolter on the floor above into five grades: bran, middlings, white shorts, second-grade flour for resifting, and fine white bread flour.

Overleaf: In the tide-powered Saddle Rock Grist Mill, the miller's office shares the second floor with four run of grindstones. Tallying his records in his ledger with a quill pen, he relies on nineteenth-century newspapers and a little whisky to keep out the chilly Atlantic air.

4

THE DISCOVERY
OF INDEPENDENCE

Helen Fox

4

YANKEE INGENUITY

In 1660 the colonists were still transplanted Europeans, looking back to their homelands for capital, supplies, customs, and manners. Europeans, yet far from typical. They had learned new skills and discovered old ones. They had survived Indian attacks or, more often than is recognized, had learned to live and work with their Indian neighbours. Not yet independent, the settlers were self-dependent. In the British colonies, Yankee ingenuity was devising ways to save time and labour. As Michel Chevalier remarked, the American developed "a mechanic in his soul." It was a quality well suited to all kinds of milling.

Early leaders built mills to benefit themselves as well as their communities. Nicholas Easton, first settler of Newport and later governor of Rhode Island, put up the first windmill in the state in 1663. A much more elaborate windmill was built in 1675 by another Rhode Island governor, Benedict Arnold, ancestor of the officer who betrayed Washington's army. It may have been modelled on Chesterton Windmill in Warwickshire, designed by a pupil of Inigo Jones. Now only its massive arched stone base still stands in Touro Park at Newport.

William Penn was part owner of the first sawmill in Pennsylvania, the colony granted him in 1681 by Charles II in payment of a debt of £16,000. As the son of a rich admiral, Penn had been a playmate of the Stuart princes; as a Quaker, he had been imprisoned in the Tower of London. Now he owned a tract almost as big as England. He wanted territory for the Society of Friends, but he also wanted to sell land to industrious families of all faiths and nationalities. He advertised for German settlers, the best farmers America had yet seen. He recognized the rights of the Indians; Cornplanter, the Seneca chief, had his own sawmill in Warren County. Though he spent only a few years in his colony, Penn laid firm foundations for its future prosperity.

His capital, Philadelphia, quickly became a focus of culture, and the need for paper to print news and opinion grew urgent.

◄ *Across the water of the Pocantico River is the grist mill of Philipsburg Manor in North Tarrytown, New York. The mill and manor house behind it are restorations of those originally owned by Frederick Philipse, a flour merchant who shipped the produce of his 90,000-acre estate down the Hudson River in the early eighteenth century. Charles Howell, a fifth-generation miller from England, supervised the construction of the mill where he now grinds corn. A master millwright, he often acts as consultant for other restorations.*

Imported paper was so scarce that schoolchildren used birchbark and hornbooks. The first American paper mill was built in 1690 at Germantown, Pennsylvania, by William Rittenhouse, a Mennonite minister who had learned paper-making in the Netherlands. The Wissahickon River provided water to clean rags and drive the water wheel for beating linen to pulp and pounding the finished paper. When the mill was carried away by a freshet, Penn asked his citizens to help rebuild it. Fostered by the printers and publishers of Philadelphia, paper mills multiplied in the Middle Colonies through the eighteenth century.

In 1699, when Penn came back to America after fifteen years in England, he and two partners built a mill at Chester which they later sold to Richard Townsend, the millwright who had designed Penn's original sawmill. In the next two years Penn planned two grist watermills, the Schuylkill Mill, and Governor's Mill, a heavy stone building with a gambrel roof. Leaving Pennsylvania in 1701, never to return, Penn wrote his agent James Logan from shipboard, "Get my two mills finished, and make the most of these to my profit." But they were badly built, the millstones arrived late, the dam broke, and Logan wrote back, "These unhappy, expensive mills have cost £200, besides several other accounts upon them."

THE GROWTH OF THE REGIONS

By the eighteenth century each area was developing its own saleable crops. The south was so dependent on tobacco and rice, introduced in the South Carolina flatlands in 1668 and by 1775 exported at the rate of 100,000 barrels a year, that Georgia and South Carolina had few flour or sawmills till after the War of Independence. In the north loomed stands of ancient trees of incalculable value. Dark, ominous, surrounding, they were loathed by the first settlers who hacked and burned magnificent oaks and butternuts, but merchants soon recognized the potential of lumber and shipbuilding. In the 1770's a Maine sawmill bee was an event as roistering as a barn raising. At Bangor, "They got the mill up the first of the winter and used two puncheons and one barrel of New England rum and had not enough liquor to finish the raising and completing of the mill." The first sawmill at Calais would never have been finished without women's help: "The number of men at the raising was so small that the ladies were obliged to lend all their strength in raising the heavy timbers."

Grain was the best crop in the Middle Colonies, and New York was flour capital of British North America at the end of the seventeenth century. Then flour was often sifted not at the mill but at the bakery, and in 1678 New York was granted a monopoly on bolting flour within the province. Mills sprang up and flour manufacture so quickly became the city's chief source of income that a new seal was adopted in 1686, adding windmill and flour barrels to the beaver as emblems of industry. Trading flour to the West Indies for rum, New York increased city revenue from £2,000 to £5,000 within seven years.

THE FORTUNES OF NEW FRANCE

Furs and flour were the currency of New France as well, though few officials recognized their priorities till the beaver market collapsed in 1705. Louis XIV was more farsighted. In 1663 he revoked the charter of the Company of One Hundred Associates and took the business of colonization into his own hands. In five years he spent a million livres on cash grants and free transportation for emigrants. Between 1666 and 1668 the population of New France doubled to six thousand. The settlement on the St. Lawrence River grew in substance, though in a very different spirit from that of its British neighbours. Here were no colonial assemblies, town meetings, or self-made merchants. To the end of the regime, the colony would be ruled from France, subject to each king's whims and his wars in Europe, burdened with too many officials whose authority overlapped, perpetually begging France for money and supplies. But Louis made a promising start by appointing as his first intendant Jean Talon, the perfect executor for his plan.

Talon forfeited unclaimed seigneuries and awarded them to abler aristocrats. He sent a regiment to keep Mohawks and English out of the Richelieu Valley, where they built fortified stone windmills to produce flour and lumber to ship to Quebec. He established industries: fishing, lumbering, shipbuilding, brewing, pitch, potash, growing wheat, peas, hops and barley for food, flax and hemp for spinning and weaving, and livestock for food, wool, and leather. By 1671 he boasted that he could clothe himself entirely in Canadien garments, and New France was exporting her surplus crops to France, Acadia, and the Antilles. Eager to ship timber, he prodded seigneurs, religious orders, and habitants to build sawmills; New France had six by 1717 and fifty-two by 1734.

Flour milling, always best practised by the Church, was encouraged by the Sulpicians who drew up strict detailed rules

96

for running their mills. In 1749 the Swedish naturalist Peter Kalm examined a Sulpician mill on Montreal Island where the order had a monopoly: "The mill is built of stone with three water wheels and three pairs of stones. I noticed first that the wheels and axles were made of white oak; secondly that the cogs in the wheel and other parts were made either of the sugar maple or of *bois dur*, because that was considered the hardest wood there; third, that the millstones had come from France and consisted of a conglomerate and quartz grains, both of the size of hazelnuts and ordinary sand, all bound together by white limestone."

Kalm, whose meticulous observations extended far beyond his own field of botany, looked for mills everywhere. Impressed by the water power of Cohoes Falls, he tested Albany well water and found it "inhabited by an abundance of small crustaceans" which swam briskly even when he added a quarter-part rum, suggesting that "this water is not very wholesome for people who are not used to it." In New France he saw water and windmills along the St. Lawrence. The windmills, stone towers with wooden caps, generally had thin board sails, but two outside Montreal had linen sails which were removed after grinding. Kalm also took time to note that French women wore very short skirts and very high heels but worked much harder than "the English women in the plantations" of the south.

Invited to the seigneury of the governor of Montreal, Baron de Longueuil, he found the governor's flour mill on Ile Madeleine driven by the swift St. Lawrence current without need for a dam. The baron, Charles le Moyne, was the son of one of the few seigneurs who had settled early at Ville Marie, building a huge chateau, a church, a brewery, and a fine mill. He and his ten brothers, audacious soldiers and explorers, were the heroes of New France.

By Kalm's day, a series of edicts had forced New France to improve flour production. Seigneurs were again ordered to build banal mills; sieves were issued so that only clean wheat was milled for export; and millers were punished for giving short weight or taking brandy bribes from rich customers who wanted their grain ground first. In 1689 a miller found guilty of having stolen wheat was condemned to be hanged beside his mill. On appeal, he was sentenced to be beaten and branded with the *fleur de lys*, but the Sulpicians whose mill he rented won him a pardon. The settlers who manned Fort Pontchartrain, established in 1734 to protect the French trading empire, were required to build windmills for flour and defence along the Detroit River. By 1750 New

◀ *Encouraged by the citizens of Papineauville, Quebec, the provincial government recently bought the old flour mill for restoration. The long flume from the dam behind the mill carries water to the turbine under the mill.*

Helen Fox

98

France had 150 flour mills and a sizable export to the Antilles.

Expansion to Detroit began in a period of prosperity, when young Louis XV and Intendant Hocquart shared high hopes for New France. In 1733 they backed ambitious plans for an ironworks at Saint-Maurice near Trois Rivières. Everything went wrong. The big forge needed six waterwheels but the stream had only enough water for two. Forest fires diminished the wood supply. The first owner died; the second went bankrupt; the third owed so much to the French government that the exasperated king took over the ironworks. In the 1740's four forges made pig iron, pots, stoves, nails, weapons – bombs, cannonballs, musket barrels, cannon for forts and warships – and a seven-year profit of £72,286, but debt and bad management still jeopardized the operation.

This puzzled Kalm: "Here are many officers and overseers who have very good houses, built on purpose for them. It is agreed on all sides that the revenue of the ironworks do not pay the expenses which the king must every year have for maintaining them. They lay the fault on the bad state of the population and say that the inhabitants in the country are few, and that these have enough to do in attending to their agriculture, and that it therefore costs large sums to get a sufficient number of workmen. But however plausible this may appear, yet it is surprising that the king should be a loser in carrying on this work, for the ore is easily broken, very near the furnaces and very fusible. The iron is good, and can be very conveniently transported over the country. These are, moreover, the only ironworks in the country from which everybody must supply himself with iron tools and what other iron he wants."

Astute as he was, Kalm failed to understand how the precarious existence of New France fluctuated with the fortunes of the homeland. What Talon and Hocquart had accomplished was almost obliterated by Bigot, the last intendant, who forbade the export of lumber, allowed only his favourite millers to grind flour and traded in wheat for his own profit.

France had tried valiantly to hold her North American colonies. In 1690 Quebec had repelled New Englanders led by Sir William Phips. In 1754 Washington's expeditions against French forts in the Ohio valley had been defeated. Even in Acadia, constantly changing hands between French and English, the French had survived. While the English held Port Royal in 1693, Governor Robineau de Villebon set up Fort Saint Joseph north of Saint John, where the Sieur de Chauffours had built the first saw-

◀ Typical of the seigneurial grist mills of New France, le Moulin de Montarville at Mont-Saint-Bruno was built of wood by Pierre Boucher in 1710 and rebuilt in fieldstone by René Boucher de la Bruère in 1741. When the seigneury was sold to Drummond, Pease and Birks in 1897, the mill was turned into a chapel. Now the estate is a provincial park, and it is hoped that the only banal watermill left in the region may someday be set to work at its original task of grinding grain.

mill in New Brunswick. Near Tatamagouche, French settlers had mills and a primitive copper-smelting works. William Pepperell from Maine took Louisbourg in 1745, but to the chagrin of the New Englanders Great Britain traded it back to France for Madras three years later.

When the English scattered the Acadians along the Atlantic seaboard in 1755, two thousand of them found refuge with Sieur Nicholas Gauthier in Prince Edward Island. The richest man in Acadia, Gauthier had lived forty years at Port Royal where he owned two farms, two flour mills, a sawmill, and two ships for trading with Louisbourg, Boston, and the West Indies. Having broken his vow of neutrality in an attack on Annapolis Royal, he moved to Prince Edward Island and re-established himself at Bel-Air on the Northeast River.

With the outbreak of the Seven Years' War in 1756, the tide turned against France. The British took Fort Detroit, Fort Duquesne at Pittsburgh, Fort Frontenac at Kingston, Louisbourg, and, in 1759, Quebec. When the French regime ended, New France had fifty thousand citizens and eight million arpents of land, a quarter of it owned by the Church. General Murray had a hot temper. He ordered miller Nadeau de Saint-Michel, who supplied flour to the French army after the battle of the Plains of Abraham, hanged from his mill and his corpse exposed three days. But Murray's surrender terms allowed the French to keep their faith and customs, and his successor as governor, Guy Carleton, confirmed their rights by the Quebec Act of 1774.

The English who ruled by law inherited a system ruled by custom, and isolation from France frayed the ties between seigneurs and habitants. Some seigneurs sold their land to British officers and returned to France. Now under the control of an aggressive commercial nation with a relatively consistent colonial policy, Canada expanded her trade, agriculture, transport and communication systems. Britain followed the pattern she had set for the colonies to the south, stimulating industries from which she could profit and suppressing those that competed with her home products. When Carleton sent word that habitants who could not afford imported clothes were making their own, the government tried to prohibit textile manufacture.

Flour, on the other hand, was a useful export, and milling was revived and improved. In 1787 Alexander Davison reported on Quebec to the Office of Trade of Plantations: "The Flour indeed that was and is made in the Province by the old French Mills might be exceptionable and unprepared for keeping – therefore

Colonial commander William Peppe-▶ rell rides out in triumph from the smoking ruins of the French fortress of Louisbourg on Cape Breton Island. Since the late seventeenth century the Pepperell family had owned vast estates stretching from the Piscataqua River to the Saco River in Maine. Along these rivers they built saw and grist mills that yielded handsome profits.

1

2

3

4

Franklin's view. The Stamp Act of 1765 prompted his declaration, "The sun of liberty is set, you must light up the lamps of industry and economy." The act, which taxed not only newspapers and pamphlets but legal, commercial, and other documents, was designed to reimburse Britain for her protection of the colonies during the Seven Years' War, which had left her the most powerful nation on earth but ruinously in debt. But it was also planned to hamper ordinary colonial business. Though, like earlier restraints, it was often violated in practice, it spurred the colonies to such vigorous protests against trade with Britain that English merchants lost money and English workers lost jobs. Applied to Canada as well, the Stamp Act prompted businessmen to complain, "We are no less sensible than our Neighbours" of its "ill Consequences," and to petition for an assembly to control taxation. The act was repealed, and other duties imposed in 1767, leaving only a token tax on tea. But by now even a token was too much.

As events built toward revolution, patriotism gave new impetus to home manufacture. Tea and mutton disappeared from tables; sheep were saved for their wool. At funerals, mourners wore crepe armbands instead of imported black clothes. Young ladies of fine family became "Daughters of Liberty" who set an example by holding daylong spinning matches. In 1768 the graduating class of Cambridge College, later Harvard, declared their sympathies by wearing homespun. George Washington, accustomed to ordering scarlet coats, ruffled shirts, and gold lace from his London tailor with instructions, "Whatever goods you send me, let them be fashionable," would wear a dark brown suit of wool made in a Connecticut mill for his inauguration in 1789.

With the Declaration of Independence, the mills of the United States came into their own. They had always been useful. Now they were essential. They had to produce everyday necessities whose import from Britain was suddenly forbidden, whose purchase from other countries was beyond American means – and, for the first time, they had to supply the tools of war. Throughout the revolution, their strategic importance was recognized in raids, defence plans, and anguished military dispatches demanding their products or deploring their destruction. Without the existence and expansion of mills of all kinds, the colonists and the Continental Army could not have survived the first three years of war when the Thirteen Colonies fought alone against an adversary so formidable that only the most optimistic revolutionaries were convinced they could win.

George Washington at his inauguration in New York City.

The present sawmill in the historic ironworking village of Old Batsto, New Jersey, was built in 1882 with a circular saw powered by a turbine. During the War of Independence and the War of 1812, when the town became a vital source of weapons, the saw used was an up-and-down type built by John Fort in 1761.

THE SAWMILL

Lumber for the Pioneers

Of all mills, sawmills are the least romantic, the most conspicuously functional. The legends of lumbering were spun round the roistering loggers who hurtled downriver in spring on rafts made from the timbers they felled. By comparison, sawing boards, planks, and deals seems prosaic. The prettiest thing about a sawmill is usually the millpond.

Yet a good sawmill has its own character, austerely practical, supremely efficient. The power transmitted from turbine or diesel engine emerges instantly in the naked blade ripping through hardwood. Only the saw is fixed; everything else in the mill is tuned to its speed. Its setting governs the rate at which the log carriage feeds it lumber, and the movements of the men who load and unload the logs on the carriage are as rhythmical as ballet steps.

Today, sawmills with reciprocating saws driven by vertical water wheels are found only in restored or reconstructed pioneer villages. Commercial mills use circular saws, and only a few are still powered by turbines. But these small country mills, producing lumber for local houses and barns, are closer in spirit to the mills of early settlers than to the giant water-powered nineteenth-century sawmill complexes that shipped wood to Europe and the prairies. Some still do custom work, sawing logs from a farmer's woodlot into dressed lumber of precisely the right dimensions to build his new house or barn.

2

1. Though it has stood for 150 years and was working not long ago, Adam's Sawmill near Dewitt Corners, Ontario, has already acquired the temporary, derelict look of an abandoned mill. Carl Adams, who bought it in 1913, stretched the flimsy board dam farther and farther to increase his power by enlarging a shallow millpond.

2. Now the log carriage is idle. Adam's Mill was one of the last in the province to run on water power. When John Ritchie owned it in 1900 he employed thirty men.

1

2   3

4

The Lost Timber Country

Immigrants who sailed to Canada or the northeastern states packed in the hulls of homeward-bound timber ships would scarcely recognize the landscape now. Then it was covered with forest, so dark and dense that it seemed impenetrable. More than any other industry, the lumber trade has changed the face of our land. Over the past two centuries, thrusting farther and farther into the backwoods, it has gradually stripped the forests.

Nineteenth-century sawmilling was a prodigal process that dumped millions of cubic feet of slabs, bark, edgings, and sawdust into rivers every year, killing fish and gradually blocking shipping. In 1873 a Canadian commission on the condition of waterways reported that sawmill

waste had filled all the small harbours and creeks on the Miramichi and reduced some reaches of the Ottawa River from up to 180 feet to depths of a few feet.

1. The mill on Little River in East Lebanon, Maine, was built in 1774 as a community mill, each man helping to raise it and running it for a day when his turn came around. Now a National Historic Site, it is being restored to pre-Civil War running condition.

2. Darr's Mill at North Wentworth, Nova Scotia, once split shingles with power from a turbine fed by a dam across the stream.

3. Through an 850-foot canal dug by hand in 1826, water flows into the millpond at McDonald Brothers' Mill, reconstructed in Sherbrooke Village, Nova Scotia. Pine logs stored in the pond during the summer are "cured" by the water which draws out the pitch, leaving the wood lighter and much easier to saw or plane.

4. Mills often slip gently from one form to another. The mill of St. Isidore Dorchester was built as a seigneurial grist mill early in the nineteenth century. When Damase Parent bought it in 1892, it was a flour and sawmill. Now his grandson runs it as a saw and planing mill.

More than most sawmills, Masse Lumber Company has an air of permanence. Its two substantial buildings lie west of the millpond dam in East Vasselboro, Maine. On the left, the sawmill reverberates with whirling, whining saws, chains, and belts, the throb of diesel power, and the ceaseless flow of water. Its floor boards are hand-hewn timbers a foot square, fifty-seven feet long, and nearly two hundred years old. On the right, the old grist mill now houses a machine shop and carpentry workshop where the equipment to maintain and repair the sawmill still runs on power from a thirty-five horsepower scroll turbine.

Until 1912, the two mills had separate owners locked in a chronic wrangle over water rights. Louis Masse solved the problem by buying both mills. At seventeen, speaking no English, he had come from Quebec to work as a sawmill hand. At thirty-six he had acquired an education, a wife, three children, and a local reputation as a handy carpenter and millwright. By 1914 he had installed water systems in the two villages of China and East Vasselboro.

His son Herman took over the sawmill in 1926 and bought it from his father three years later. Now Herman's son Kenneth manages the mill and his grandson Matthew works there.

1

2

3

4

The Machinery of Sawmilling

1. At the Batsto Sawmill, the sluicegates work with a rack and pinion mechanism adjusted by putting a pole into the lever wheel.

2. The control lever of the planing machine in the Little River Sawmill at East Lebanon, Maine.

3. A log trolley stands idle at Krug's Sawmill in Chesley, Ontario, burned out in 1974.

4. At Adamson's Sawmill in York Mills, New Brunswick, a conveyor loads sawdust into a hopper above the truck.

5. The two men are canting a log to be placed on the saw carriage in the foreground at Upper Canada Village, a pre-Confederation community reconstructed at Morrisburg, Ontario.

6. With each stroke of the reciprocating blade of the up-and-down saw at Upper Canada Village, a ratchet on a gear beside the log carriage inches the log toward the saw.

7. At Sherbrooke Village, Nova Scotia, the sawyer adjusts the log for the next cut. A black spruce high breast wheel, twelve feet in diameter, powers both the saw and the log carriage. The mill, a reconstruction designed and built by local millwright Peter Breen, is on the site of the mill owned from 1856 by Alexander and David McDonald, sawmillers and shipbuilders who sent barques up to 749 tons sailing down the St. Mary River.

Overleaf: At the Masse Mill in East Vasselboro, logs are stripped of bark by a machine made from an old farm tractor.

5

6

7

5

PATRIOTS
AND LOYALISTS

Helen Fox

5

FIGHTING FOR INDEPENDENCE

From the day that Britain closed the port of Boston to trade in 1774, uniting the colonies in support of Massachusetts, American manufacturing became not a patriotic gesture but the means of survival. The war thus gave enormous thrust to milling. The rebels were short of everything – gunpowder, food, clothing, money, paper. The Journal of the Second Session of the Assembly in 1781 went unpublished because New York had no paper on which to print it.

The ill-matched battle of Bunker Hill was an omen of things to come. Wars with French and Indians had turned Americans into a nation of militiamen whose methods British regulars considered unsporting. Half the redcoats who stormed the hill (not Bunker but Breed's, lower and less easily defended) in close ranks, burdened with 120-pound packs of rations and blankets, lay dead or wounded before the colonists fled – because they had run out of gunpowder. The first powder mill had been built at Pecoit, Massachusetts, in 1639, but little powder was produced in the colonies until England prohibited its export in October 1774. Powder mills hastily set up in seven colonies were profitable and productive (and highly explosive), but the Continental Army had to depend on powder from Europe and the West Indies as well.

The army's chief source of iron for weapons, the Salisbury district of Connecticut, was supplemented by ironworks in Pennsylvania, New Jersey, Virginia, and Maryland. All the southern furnaces and some northern ones were manned by slaves. Ore for the Principio furnace on Chesapeake Bay was hauled fifty miles in galleys rowed by white convicts. Locksmiths and blacksmiths made firearms. Eli Whitney pounded out nails by hand; in 1798, after his invention of the cotton gin had brought him no profit, he would become the government's most ingenious maker of arms.

The shortage of clothing was desperate. In the bitter winter of 1777-78, more than 2,500 of Washington's troops at Valley Forge died and 2,000 deserted. Of his remaining 6,000 men, half lacked shoes and warm clothing. Samuel Wetherill, Jr., the Philadelphia

◀ *Built a century ago as a woollen mill, Falls Mill in Belvidere, Tennessee, now uses a thirty-four-foot water wheel to produce stone-ground cornmeal and flour.*

wool and cotton manufacturer who formed the "Fighting Quakers," was credited with helping to keep the remnant from disbanding by supplying them with clothes.

When France, which had been secretly sending munitions, openly joined the rebels in 1778, shabby American troops shrank from appearing beside their splendidly uniformed allies. Soon the French were shivering, too. At a ball in Baltimore, General La Fayette rebuked his hostesses: "You are very handsome, you dance very prettily; but my soldiers have no *shirts.*" The music stopped and the ladies went home to sew war clothes.

The French alliance revived American hopes of enlisting Canada in the revolution. It should have been easy. Quebec's 90,000 citizens, mainly French, were still chafing under recent conquest. Three-quarters of Nova Scotia's 20,000 had come from New England, many within the past ten years. Sympathizers in both provinces actively helped the rebels, but Canada resisted being drawn into a conflict that was not of her own making, and whose results were still unpredictable.

France's greatest contribution to the Americans was a subsidy of six million livres and security for a loan of ten million from the Netherlands, where English cloth was bought and shipped to America. Gold and silver currency had almost vanished, and the value of paper money dropped so fast that by 1780 army suppliers were refusing to produce on credit.

Money, manpower, clothing, food – as the war raged, Congress juggled shortages and patriots responded. Victory hung in the balance. Wheat from Pennsylvania was sold to Havana for millions of dollars. Only one man from each mill was excused from army duty, and women in the Middle Colonies helped grind flour for troops, refugees, and residents. The mills of the Schoharie valley shipped flour to Gates and Schuyler as Burgoyne moved ponderously down to defeat at Saratoga in 1777. At Chrysler's Hook, Loyalist Adam Chrysler ordered Indians to burn his grist mill to keep it from falling into rebel hands.

From New England to North Carolina the British seized or destroyed mills. The most vulnerable, in upstate New York, were guarded night and day against Loyalist regiments like the Royal Greens of Sir John Johnson. On his most devastating raid, in the fall of 1780, Sir John led six hundred men from his own regiment, Butler's Rangers, Six Nations Indians, and British regulars, who scourged the valleys of the Schoharie and the Mohawk. In three days his troops burned thirteen grist mills, several sawmills, and

Loyalist Sir John Johnson sacked and burned the Mohawk Valley, where his own estate had been confiscated by Congress.

a thousand houses with their barns holding grain and fodder desperately needed for the army.

As the peace treaty was signed in Paris in September 1783, American soldiers straggled home starving, in rags, unpaid but victorious. At the outbreak of revolution John Adams had reckoned that, of the Thirteen Colonies' three million citizens, about a third were rebels, a third uncommitted, and a third Loyalists. At its end, there were only winners and losers in a civil war that split families and friendships. The Loyalists who had allied themselves with the British were the losers, and Washington remarked that he "could see nothing better for them than to commit suicide."

Instead they quit the States. While the most prosperous sailed for Great Britain, about forty thousand were evacuated by ship to the Maritimes or fled across the border at Niagara and Detroit. From the main refugee shelter at Shelburne, Nova Scotia, poorly chosen and soon abandoned, most moved to New Brunswick where they built first windmills, then watermills. Charlotte County had eight sawmills in 1785 and twenty-three by 1803. The biggest was Colin Campbell's Brisk Mill which produced three million board feet a year. In 1795 the first grist windmill in Saint John, later converted to a poorhouse, was built where the Admiral Beatty Hotel now stands.

POWDERWORKS ON THE BRANDYWINE

Eleuthère Irénée Du Pont came to the United States to escape the violence of the French Revolution and stayed to become the most successful manufacturer of gunpowder in America. In 1802 Du Pont built a factory for "Military and Sporting Powder" on the Brandywine River in Delaware. The first black powder produced by his Eleutherian Mills in 1804 was so much better than any then made by Americans that his sales trebled in a year. His production kept pace with the nation's expansion westward. His powder supplied frontiersmen still dependent on game, engineers blasting out roads, canals, and mines, and the army and navy. Once deeply in debt to his French backers, he died in 1834 leaving assets worth $300,000 and a business that grew to a family fortune.

1. The rolling mills housed the most dangerous step of the powdermaking process, blending the raw ingredients.

2. Cast-iron edge runner stones crushed and mixed refined saltpetre, sublimated sulphur, and charcoal to make gunpowder.

3. All tools were wooden; a metal shovel could spark an explosion.

4. Four gates control water flow through a mill dam. The outer gates are open, with flume covers raised; the covers of the middle gates are down.

5. A sixteen-foot water wheel between the two Birkenhead rolling mills drives edge runner stones in both mills.

6. By mid-nineteenth century the wooden wheels had been replaced by turbines. The flume above the wheel feeds water to the sunken turbine.

7. From the 1830's, work cars on a narrow-gauge railway carried powder and equipment through the mill yards.

1

2

3

4

5

6

7

121

A view from inside one of the rolling mills at Du Pont's powderworks shows how the mills were designed to minimize damage if the gunpowder ingredients exploded. Three walls were built of heavy masonry to contain the blast. The back of the mill was open so that the explosion would be vented harmlessly to the river behind. The light roof, of which only the beams remain, was a second precaution, planned to direct the explosion upwards, leaving the main walls and neighbouring mills undamaged.

LOYALTIES IN CONFLICT

Denounced as traitors in the United States, the Loyalists were revered as forefathers in English Canada. Great Britain gave them free land and about £60,000 worth of transport, food, clothing, shelter, and farm tools. Refugees and disbanded troops settled by Governor Haldimand in the Quinte area along the north shore of the St. Lawrence River and Lake Ontario were issued hand grist mills until the first government mill, which ground toll free, was built at Kingston Mills in 1783 by Robert Clark. A New York millwright who had fought with the British Army and Jessup's Rangers, Clark saw his wife and two children for the first time in seven years when they reached Cataraqui in 1784. In 1787 he completed a sawmill and grist mill on the falls at Napanee. Soon Loyalists in each town were building mills: John Meyers at Belleville, Peter van Alstine at Adolphustown, Henry Ripson at Trenton. At Fredericksburg, John Howell built a windmill. Sir John Johnson, who had directed settlement of the Quinte area in 1784, built the first mills at Williamstown in Glengarry Township.

In 1783 Sergeant Brass of the Royal Greens was dispatched to build grist and sawmills at Niagara, and others followed: Green's Mills at Grimsby, John Backhouse's mill at Port Rowan, the complex of mills begun by John and George Ball at Ball's Falls, and mills along the Lake Erie shore built by Loyalists from the Maritimes. Captain Samuel Ryerse's mills at Long Point, though badly built, were the only ones within a seventy-mile radius. By 1792, when the Niagara region had nineteen grist and thirteen sawmills, Upper Canadian Loyalists were selling their surplus wheat and flour to the garrisons at Kingston, York, and Niagara.

The Eastern Townships of Quebec present a paradox. Here, where Loyalist tradition is strong, the first official Loyalist grant was not made until 1796, when about 25,000 Americans had already moved to the province. Haldimand carefully settled his new citizens away from a border where sympathies were uncertain. The first settlers of the Townships, some of them squatters, were a mixture of Loyalists, French, British immigrants and troops, and Americans attracted by land granted freehold without seigneurial dues. They were an industrious lot who quickly tired of hand grinding and hewing. Again, towns like Shipton, Potton, Danville, Huntingville, Windsor, Ulverton, and Frelighsburg grew round early mills. Sherbrooke, founded in 1794 by Vermont Loyalist Gilbert Hyatt, was called Hyatt's Mills for the next fifty-eight years.

Kingston Mills, built by the Canadian government for the Loyalists in 1893, is shown in a detail from a watercolour by an unknown artist.

As recently as 1962 the old Barrington ▶ Woolen Mill still produced lobstermen's mittens, gargantuan gloves soon shrunken, matted and yellowed from handling bait. Early settlers from Cape Cod and Nantucket had found that the rocky headlands and islands of Nova Scotia's southwest tip provided good water for fishing and land for grazing sheep. The low extension, where fleeces were washed with homemade soap, is floored with pebbles from the Barrington River that powered the mill for nearly eighty years.

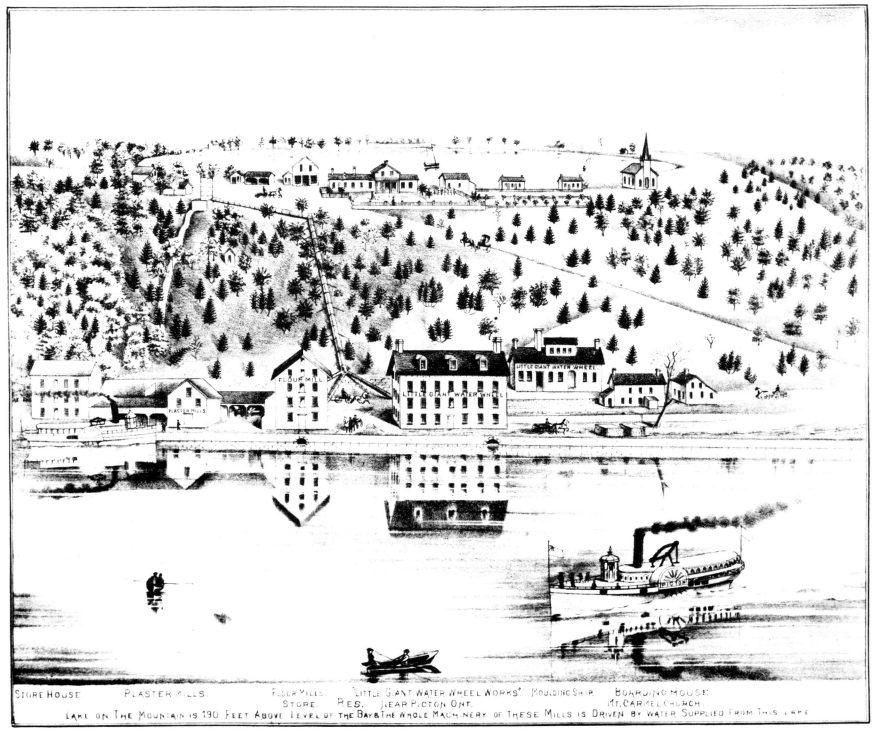

STORE HOUSE   PLASTER MILLS.   FLOUR MILLS.   "LITTLE GIANT WATER WHEEL WORKS". MOULDING SHOP.   BOARDING HOUSE.
                                          STORE.   RES.   NEAR PICTON ONT.   MT. CARMEL CHURCH.
LAKE ON THE MOUNTAIN IS 190 FEET ABOVE LEVEL OF THE BAY & THE WHOLE MACHINERY OF THESE MILLS IS DRIVEN BY WATER SUPPLIED FROM THIS LAKE

At Glenora Mills near Picton, Ontario,
a flume from the Lake on the Mountain split to drive three mills. The
biggest mill building is now used as a
government fish hatchery.

The grist mill at Ball's Falls, Ontario,
hand hewn by George and John Ball in
1809 and fitted with four run of
stones, was driven by a thirty-two-foot
overshot wheel on the Twenty Mile
Creek.

British sympathizers were still crossing the Quebec border through the early nineteenth century. The Cornell Mill, now the Missisquoi County Museum at Stanbridge East, was built in 1830 by Zebulon Cornell from New York State, whose cousin Ezra founded Cornell University. Paul Holland Knowlton bought nearly half a township of wild land at Coldbrook in 1834 and put up saw and grist mills within two years. In 1837 a traveller wrote, "Mr. Knoulton's (*sic*) improvements directly yield a handsome return even at this early period of their existence. His grist-mill in particular, from its superior excellence, drains at least twice as extensive a district as Mr. Knoulton's most sanguine expectations had anticipated – actually drawing business past the doors of other mills. As the machinery is in perfect order, and the supply of water has been rendered abundant by artificial means, the farmer can carry home his grain in the shape of flour with the least possible delay, whereas mills in indifferent order or occasionally deficient in power might render necessary two journeys and much loss of time.... Milling is undeniably the best business in the Eastern Townships." Knowlton's influence as community leader and member of the Legislative Council of Canada was recognized when the town was given his name in 1851.

The Loyalists brought a mixture of conservative and democratic convictions that would pervade Canada. Their arrival unleashed a tide of immigration from Europe in which French Canadians were soon outnumbered by English, Scots, and Irish. New boundaries had to be set: New Brunswick became a province, and Upper Canada, now Ontario, was separated from Lower Canada, now Quebec. Industry scarcely existed: milling of all kinds was still in the pioneer stage.

In the hothouse of revolution, the United States had already developed a system of self-government and a diversity of mill manufactures. The war left her starved for goods, forty million dollars in debt to European suppliers, and determined to establish economic as well as political independence.

The post office, linked with the ▶ Harris family since it opened in 1848, is now in a corner room in the house built for their patriarch, Bethuel Harris. His son Charles C. P. Harris, postmaster for forty years, used his parlour as his office. It's still a town meeting place where the notice board advertises free kittens, and a sale at the Harris Mill's rival, Cheshire Woolens.

MILLTOWN
Harrisville, New Hampshire

The Town
Where the Past
Lives On

Harrisville, New Hampshire, is the only nineteenth-century textile mill town that stands almost unchanged today. An accidental anachronism, it has escaped the mainstream of development that razed or swallowed bigger complexes. The style of its buildings, neo-classical rendered in red-brick vernacular, unifies the town. Their sites, scattered like jackstraws round two streams flowing from the same pond, give it agreeable variety.

The village was called Twitchell's Mills when Bethuel Harris married Deborah Twitchell and set up a carding machine in his father-in-law's grist and sawmill. By 1830, when Bethuel's son Milan built the Harris Mill, woollen manufacture engrossed the town he renamed Harrisville, and almost everyone there was a relation or employee of the Harrises.

Forceful, contentious, Milan Harris dominated the community for a generation. His influence waned from mid-century, when his brother Cyrus died of consumption. Cyrus's handsome new mill of hewn granite, farther down the turbulent torrent of Goose Brook than Milan's mill, was sold to the Colony family. Rivalry between the Harris Mill and the Cheshire Mill began with a bitter dispute over water rights and flared into bizarre gunfire after the presidential election of 1856. The Harrises were Republicans and rock-hard conservatives. The Colonys were Democrats in politics and progressive in outlook. According to Milan, Timothy Colony was ringleader of a "rabble" that celebrated the Democratic victory by firing a cannon at him. Milan was rough-handled, his mill windows were shattered, and a rowdy mob broke up a prayer meeting in the Congregational Church founded and funded by the Harrises. It was clear that the town was no longer their private preserve.

The Evangelical Congregational Church is the focal point of the Harrisville landscape. Bethuel Harris paid more than half the cost of building it in 1842, and Harris donations and attendance supported it.

Industry in Harrisville

1. *Goose Brook ravine is so steep that the Harris Mill and boiler house in the foreground had to be built across it. The mill, with its eyebrow monitor lighting the attic, follows the Rhode Island style introduced by Samuel Slater. The tower and cupola, added in the 1840's, had to be built on firm ground at the end rather than their usual placement in the middle of the long side.*

2. *Downstream from the granite mill is a repair shop and, in the background, the brick mill built in 1859 to expand the production of the Cheshire Mill. As the fortunes of the Harris family declined, the Colonys ploughed back their profits into new buildings and equipment.*

3. *The loading bays indicate that this 1850 building on Main Street was once a storehouse. Early this century it was J. H. Farwell's store.*

2

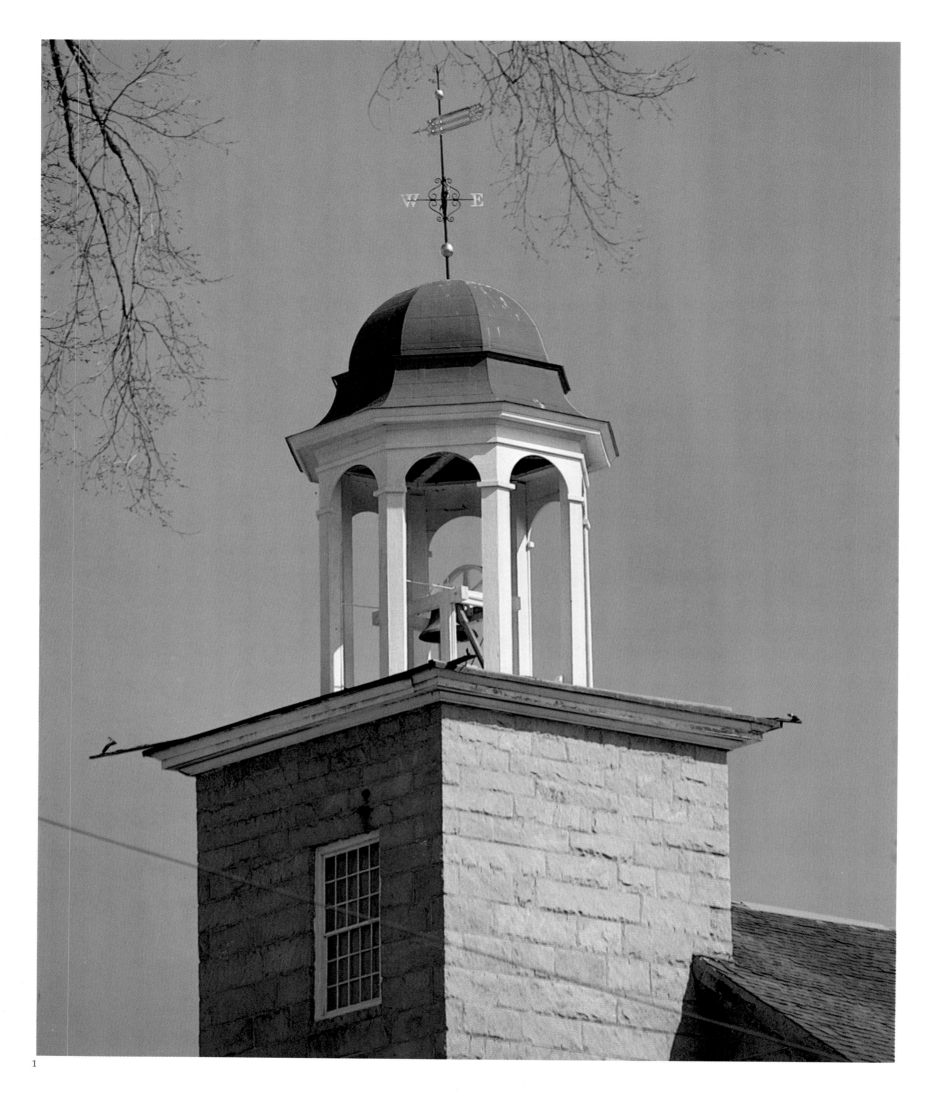

2

3

Two Rival Mills

The Civil War kept both mills busy making "Government Blue" flannel for uniforms, but Milan expanded too late. In 1867 he spent $75,000 on a new mill, and the economic slump of 1874 left him bankrupt at seventy-five. After his death the Colonys bought his mills, storage sheds, and workers' boarding house and cottages. By the turn of the century there were no Harrises living in Harrisville and only the Cheshire Mill was still working. When it closed down in 1970, the town slipped subtly from production to preservation. Fortunately, Harrisville is small enough to make restoration feasible.

1. Like the Harris Mill, the Cheshire Mill had its stair tower at the end, topped by a cupola housing the bell that called hands to work.

2. The hoist lifted bulky loads to the upper floors of the mill.

3. Milan Harris would have been chagrined if he had lived to see his sorting house renamed for his rivals.

Overleaf: The vestry overhanging the pond was originally a Congregational chapel advocated and mainly paid for by Bethuel Harris. Later it served as a schoolhouse, a chilly place where the wind whistled through inch-wide cracks between floor boards. Restored, it now houses the town library.

135

6

TWO INDUSTRIOUS REVOLUTIONARIES

6

OLIVER EVANS: INVENTOR

Before the century ended, two remarkable men sparked an explosion of new ideas and techniques that thrust the United States into the Industrial Revolution. Oliver Evans, a native American, brought automation to flour milling. Samuel Slater, an Englishman, brought mass production to textile manufacture. Each realized his system by means of the abundant water power of America.

Oliver Evans of Newport, Delaware, was a mechanical genius who, at eighteen, worked out the details of a high-pressure steam engine. He dreamed of using it for transportation: "A carriage will set out from Washington in the morning, the passengers will breakfast in Baltimore, dine at Philadelphia, and sup at New York, the same day." But rivals were more successful in developing steam transport, and Evans turned his attention to milling.

Milling was something he knew practically nothing about until he was twenty-seven, when in 1782 he contracted to build a flour mill. Like other inventors, he approached the job backwards: "I first conceived the great design of applying the power that drives the millstones to perform all the operations which were hitherto effected by manual labour." When he discovered that practically all the mill work was done by hand, he was astonished and inspired. By 1785 his automatic mill, powered by an overshot water wheel, was clattering into operation on Red Clay Creek north of Newport.

Evans' mill incorporated three devices, which he acknowledged as adaptations of earlier ideas to one integrated process: the elevator, an endless band with wooden or sheet metal buckets spaced a foot apart, to lift three hundred bushels of grain per hour from ground to grain loft, from which it was fed down to the millstones; the hopper boy, a twelve-foot revolving rake that spread the ground meal evenly on the loft floor, cooling it and guiding it to a central chute over the hopper that passed it to the bolter; and the conveyor, an auger of sheet iron wound spirally round a wooden shaft encased in a wooden trough, which carried unground grain horizontally through the mill, cleaning it with a fan as it moved.

Oliver Evans

◀ *The imposing Graue Mill in Oak Brook, Illinois, served as a way station for escaped slaves following the drinking gourd – the Big Dipper that marked the northbound trail of the Underground Railroad before the Civil War. From 1852 to 1920 the grist mill was run by three generations of Graues. Restored in 1950, the mill is driven by the only working water wheel in the state.*

The essence of Evans' ingenuity was the use of endless belts, buckets, and screws to provide continuous motion without hand labour. Water power lifted grain or grist for the next stage of milling. Earlier millers had hauled their grain up on rope pulleys by hand or by water-driven hoists that robbed power from the grindstones. Evans' elevators allowed a light, continuous flow of power coordinated with smooth grinding. The same water power ran machines that moved the grain to chutes through which it descended by gravity. Evans suggested a variety of arrangements by which conveyors, chutes, screws, and movable crane spouts carried grain from process to process according to the floor layout of existing mills. All operations became streamlined, balanced, and flexible.

Evans had no instant success. Millers were already thriving; Jefferson wrote, "Virginia, Maryland, Pennsylvania, Delaware, New Jersey and New York abound with large manufacturing mills for the export of flour." Evans sent his brother to more than a hundred mills, offering his ideas free to the first miller in each county who would demonstrate them; only Jonathan Ellicott of Maryland accepted. In 1792 Evans sold his interest in his own mill and moved to Philadelphia where he was reduced to such poverty that his wife sold the cloth she had made for their children's clothes to buy bread.

His friend Robert Leslie tried to promote Evans' models in England. He found British machinery advanced far beyond American, especially in the use of steam power. James Watt's engines had been running eight pair of stones in the Albion Mills in London since 1784, and cotton machinery in Castle Mill in Nottinghamshire since 1785. But Leslie could not interest English millers in automation. He wrote Evans, "Nine-tenths of all the wheat in England is ground by little footy windmills that do not look worth 20 pound each, but the number of them is so great as to be able to do all the business, but the owners of them would sooner jog on in the old way than give five shillings for all the improvements in the world ... so great is the difference between this country and America in the art of milling, which very much surprises me, as it is the only thing I know of that is not done in greater perfection than in America." Not till the nineteenth century would Evans' ideas be adopted in Britain.

Meanwhile Evans was writing what would become a classic handbook, *The Young Mill-wright and Miller's Guide*. He had it printed and sold it himself; it ran through fifteen editions from 1795 to 1850. It was a service manual, encompassing not only

This side elevation from The Young Mill-wright *illustrates Oliver Evans' automatic mill at work. Essentially the principles Evans suggests are simple. Elevators, endless bands fitted with small buckets, lift grain up one or more storeys. Chutes, straight or slanted, direct its descent by gravity. Augers (B, C, R, and K) carry it horizontally across the mill floors. Evans' idea of unloading grain directly from the ship at left was put into practice only after his death. His lettering traces the progress of grain through each milling process.*

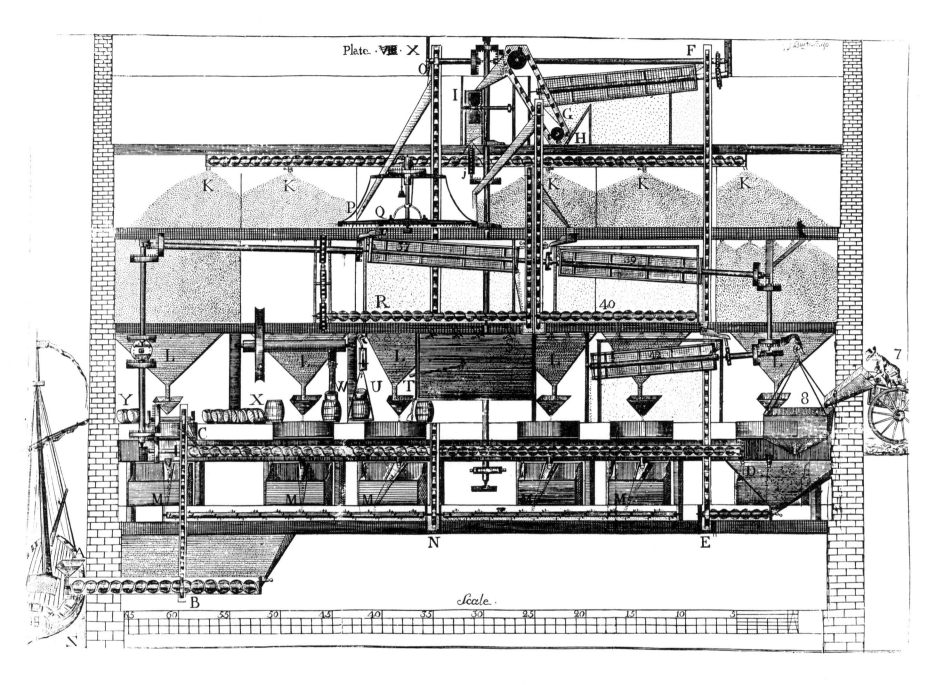

Plate VIII X

Scale.

The immense facade of Harmony Mills stretches like a fortress along the Mohawk River below Cohoes Falls and just above the junction of the Mohawk and the Hudson. The buildings to the right are boarding houses for workers in the vast textile complex.

automation but advice on every phase of flour milling, with short chapters on sawmills and fulling mills. What he had designed was in fact the forerunner of a modern plant.

His busy mind constantly devised ways of getting more fine flour from a bushel of wheat. The first gain was obvious: automation reduced loss from spilling and from dirty hands and boots. Next he used water power to clean the grain several times. He blew air through it as it was unloaded. He ran it through stones widely set so that they rubbed without grinding, then through a double cylinder of two revolving wire sieves that screened out dirt and chaff. With bolters he separated fine flour from middlings, coarsely ground wheat that retained bits of its bran coating. He then reground his middlings, running the millstones slowly. He recognized that speed, pressure, and heat destroyed the life of the flour: "Flour may be reduced to the greatest degree of fineness, without injuring the quality; provided, it be done with sharp clean stones, and with little pressure." His method produced an average of 64 per cent fine flour, between two and three pounds more than could be got from a bushel of wheat by ordinary milling.

The Ellicotts of Baltimore installed his inventions in their flour mills on the Patapsco River and found they increased production by $32,500 a year and saved $4,875 in wages. They were the leading mill owners and farmers in Maryland, and their opinions carried weight. The young Frenchman Ferdinand-Marie Bayard, travelling from Maryland to Virginia in the summer of 1791, breakfasted at Ellicott's Lower Mill, smaller than those on the Patapsco but busy nevertheless. Bayard found his Quaker host too solemn for his taste: "His speech contained the honeyed tones of our nuns of old. I have never liked this silly tone of voice in men of his sect." He was more impressed by the mill. "Their mills are spacious, well-kept, and enriched by a great number of machines which make up the deficiency in manpower.... The wheat is hoisted to the top floor by means of a machine whose mechanism is concealed; there it is spread out and descends to the mill stones. The flour falls into a room below and is carried by a mechanical contrivance to a place where it is put into barrels. The barrels are lifted away and loaded on wagons by the same machine which hoists the sacks of wheat to the upper floor. ... Americans will perfect the machines which are used for practical arts, because manpower will be very costly in their country for several centuries."

That same year another traveller watched Evans' system working a mill on the Brandywine. On his first presidential tour, George Washington was visiting Joseph Tatnall to thank him for his service in grinding corn for the Continental Army. Washington owned three mills, two in Virginia and one in Pennsylvania, and he had already ordered Evans' machinery for his mill at Mount Vernon.

A working farmer, with three hundred slaves on his 10,000-acre estate, Washington visited his merchant mills at Mount Vernon every day when he was at home, and expected them to yield a profit. He also consulted Evans about replacing his miller, a skilful man, "incorrigibly addicted to drunkenness," but did not hire the substitute Evans suggested because he asked £75 a year. Washington was no puritan – he tipped his gardener $4 at Christmas "with which he may be drunk for four days and four nights" – and he kept his old miller on. As well as salary the miller got his keep: house, vegetable garden, firewood, five hundred pounds of pork a year, a cow and her feed, and permission to raise fowl for his own use. In return he had to run and repair the mill with "a smart young negro man" as assistant, and supervise a cooper's shop that employed two black men and a boy.

Presidential patronage was no guarantee of prosperity. Evans' chief income came from selling bolting cloth, plaster of Paris which he ground with a small steam engine, and burrstones, of which he was one of the first American manufacturers. In 1803 he built a steam engine for a Mississippi riverboat, which was grounded by a flood before its trial run. Evans rented the engine and boiler to a sawmill owner who found it produced a prodigious three thousand feet of boards in a twelve-hour day and allowed him to drop most of his sawyers, who struck back by setting the mill on fire three times. The third time, building and machinery were destroyed.

In 1807 Evans combined his interests by installing a steam engine in a merchant mill in Lexington, Kentucky, the first steam-powered flour mill in America. Two years later he built his own steam flour mill in Pittsburgh at a cost of $14,000.

Though he had obtained patents for his milling improvements in several states and superseded these with a federal patent in 1790, he was endlessly engrossed in lawsuits against mill owners who used his ideas without paying his fee of $40 per pair of millstones. In 1809, driven to frenzy by a judge's ruling that patent rights infringed public right, he swore he had wasted his life on unrewarded inventions. Calling his family together, he burned all his drawings, papers, and specifications to warn his children

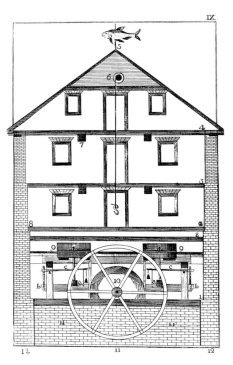

An engraving from The Young Mill-wright *shows a four-storey grist mill in which the water wheel in the foreground turns the pit wheel behind.*

against such fruitless endeavours. The mood passed but the litigation continued. In 1813, when the Quaker millers of Baltimore challenged his patent, Thomas Jefferson who for five years had been using Evans' equipment under licence in one of his mills in Virginia, supported their protests on the grounds that only the hopper boy was a new invention. Congress upheld Evans' patent, recognizing that it covered not the separate machines but their application to the milling process.

Before he died in 1819, Evans won a measure of public acceptance. Twenty-eight of his steam engines were being used for sawmills, flour mills, cotton spinning, wool cloth manufacture, drawing wire, rolling and slitting iron, papermaking, and a Boston steamboat. His flour milling improvements were widely adopted but the machinery of his time was comparatively clumsy, and few millers then had the mechanical skill to keep his potentially perfect system in smooth running order. The full impact of his ideas was realized only after his death. His methods were ideally suited for use in large merchant mills capable of processing huge quantities of grain, and the very existence of these methods made it more economical for millers to expand their operations. At the same time, other factors were combining to make large-scale production essential. As western settlement, the mechanization of farming, and the development of transport opened up the wealth of the prairies, only these bigger mills could handle the enormously increased wheat crops, and they needed Evans' equipment to do it.

One suggestion in *The Young Mill-wright* – an elevator for raising grain from a ship – was almost ignored until 1843. Then Joseph Dart, using Evans' plan as his guide, built a steam-powered grain elevator at Buffalo to replace the slow, arduous system of toting sacks of grain from wharf to warehouse on the sailors' shoulders. This speedy bulk handling gave instant impetus to the grain trade of the Great Lakes.

By mid-nineteenth century, millers were taking advantage of Evans' system of automation. By 1890, when half the flour mills in the United States had switched from water power to steam, his vision of using steam for milling at last came into its own.

The Old Slater Mill in Pawtucket is reflected in the Blackstone River above the mill dam. ▶

The mill built by Almy, Brown and Slater in 1793 measured only twenty-nine by forty-three feet. By 1835 it was enlarged by additions to west and east and the bell tower on the south side. ▶

SAMUEL SLATER'S LEGACY

ALMY·BROWN & SLATER
Cotton Manufacturers

The Machine Age Comes to Milling: The Textile Industry Revolutionized

President Andrew Jackson called Samuel Slater "the father of American manufactures." The claim was based on his successful introduction of British inventions to the textile mills of New England, but his greatest contribution was the system of mass production that has dominated North American industry ever since.

1

2

3

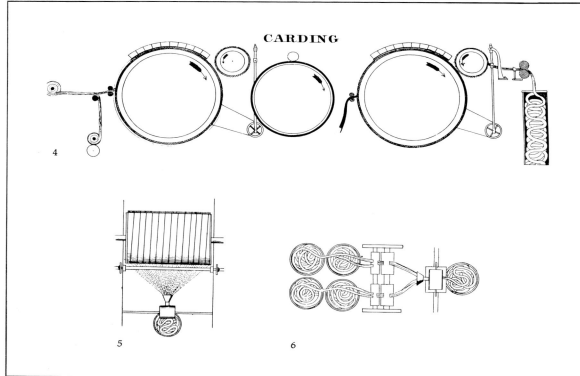

CARDING

4

5   6

1. Slater's original wooden carding machine is now in the Smithsonian Institution.

2. A model of the Arkwright spinning frame built by Slater is displayed at the Old Slater Mill in Pawtucket, Rhode Island.

3. Laps of cotton fed through the carding machine at left are combed into a thick single rope, the "sliver" coiled by the worker.

4. Each cylinder of the carding machine is set with wire teeth that comb the cotton fibres parallel. Loose cotton is fed to the large cylinder at left, stripped off by the tiny doffer above, and passed to the middle cylinder which revolves until it holds several layers of cotton – the lap. The second large cylinder and its doffer draw the lap into a sliver which is collected in a tall can and taken to the drawing frame.

5. Cotton emerges from the carding cylinder as a loose web which is pulled together into a sliver.

6. Four slivers are passed through the rollers of the drawing frame to make a tighter, stronger single thread.

1

MULE SPINNING.

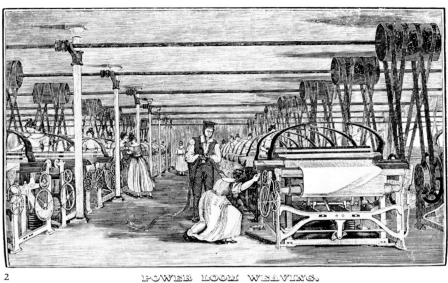

2

POWER LOOM WEAVING.

1. Invented by Samuel Crompton in 1779, the spinning mule worked swiftly and dangerously for child workers who knotted broken threads inside the moving frame on the left.

2. Power looms driven by water wheels made it possible to mechanize the entire process of textile making.

3. A power loom manufactured about 1884 weaves wool in the Barrington Mill, restored by the Nova Scotia Museum.

4. In the Barrington Woolen Mill, the skeiner, hand-turned by the iron wheel at right, produces twenty skeins of wool fed in from the twister behind, which combines strands into two-ply yarn for knitting fishermen's sweaters, socks, and mittens.

5. The carding machine from the Hapgood Mill in South Waterford, Maine, is now restored in Old Sturbridge Village, Massachusetts.

3

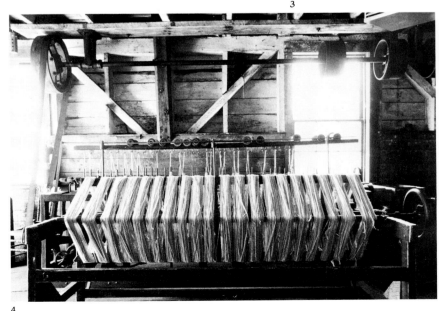

4

5

SAMUEL SLATER: INDUSTRIALIST

Samuel Slater was able to put his ideas straight to work for his own benefit. Whereas Evans was essentially an inventor, Slater was a businessman with the capital and singleness of purpose to apply the techniques of management control in his own mills. He had to. Until he established his own factory, he couldn't convince American manufacturers that mass production worked.

Samuel Slater.

Son of a rich farmer in Derbyshire, he grew up in an English farming district plunged into industry by cotton mills using Richard Arkwright's newly invented spinning frame. His father, who traded in land, arranged the sale of a mill site on the River Derwent to Arkwright's former partner Jedediah Strutt. In 1782, at fourteen, Samuel was apprenticed to Strutt, not as a mill hand but as bookkeeper and potential manager. He learned precisely how mill power was used. The power of water transferred through gears, shafts, and belts drove the machinery, and the owner's day-by-day decisions determined how smoothly the mill ran and how much cotton – and money – was produced.

Strutt's mill took advantage of inventions that had revolutionized English textile manufacture within a generation. Arkwright's spinning frame outlasted the spinning jenny of James Hargreaves because it worked on water power and produced cotton yarn strong enough to withstand the pull and friction of the loom, so that calicoes once woven with linen warp and cotton weft could be made entirely of cotton. Hargreaves and others devised a series of improvements in carding machines. In 1774 Thomas Wood's notion of fastening the cards on the cylinder spirally instead of lengthwise made endless carding possible.

The Arkwright system used by Strutt first fed handpicked raw cotton into carding cylinders mounted with bent wire spikes that combed the fibres parallel. The thick loose strand was next passed through drawing and roving frames that stretched it finer and finer, and then pulled and twisted into finished yarn by the spinning frame. Slater grasped not only the workings of each machine but the key concept that all operations must be perfectly integrated.

Slater had long planned to move to the United States when his seven-year apprenticeship ended. In England he felt the textile business was in danger of overextension by mills rapidly increasing in number and efficiency. In America he saw his chance to fill a gap created by politics and economics. Britain had jealously guarded the secrets of her most lucrative industry. Plans, models, and skilled mechanics were locked in the country by surveillance, heavy fines, and imprisonment. Americans, eager to apply their water power to their native product, cotton, were already trying to break England's hold on methods of production.

Hand spinning and weaving provided the colonies with sturdy cloth – wool, linen, and that distinctly colonial blend of the two materials, linsey-woolsey. Cotton, raw or woven, was imported and expensive until just before Slater's arrival. On the eve of revolution, the Williamsburg Convention of 1774 resolved to turn "from the cultivation of tobacco to the cultivation of such articles as may form a basis for domestic manufactures," but the southern planters lagged ten years behind. In 1784 an American ship carrying eight bags of cotton into Liverpool was seized on the grounds that the States could not have produced so much. Regular export began the next year. In 1786, when silky, long-fibred Sea Island cotton from the Bahamas was planted in the offshore islands of Georgia and South Carolina, James Madison predicted that the United States would one day become a great cotton-producing country.

Tench Coxe, grandson of the first proprietor of Carolina, had the means and influence to make this prediction come true. From 1775, when at the age of twenty he joined the United Company of Philadelphia for Promoting American Manufactures at its founding, to 1812, when his success in high government posts led Congress to ask him for a comprehensive report and recommendations on American manufactures, he was convinced that agriculture and industry should be linked. In 1787 he arranged with Jefferson, then American Minister in Paris, to send an English mechanic from Philadelphia to smuggle Arkwright models from England through France to the United States. The plan failed; English authorities seized the models and arrested the mechanic. But Coxe's advertisement in a Philadelphia paper offering a reward for English cotton machinery may have been more effective. Reprinted in England, it was credited with inducing Slater to emigrate.

Dressed as a farmer's son – which he was, and looked – and carrying only his hidden indenture papers, Slater came not as a hopeful artisan but as a prospective capitalist. He had an inheritance of $2,000 from his father's estate and a fortune in his head: a coherent theory of management and the mechanical experience to put it into practice. His timing was propitious. In May 1789, Madison had presented the first revenue bill taxing imports to protect American manufactures, including a few cotton mills struggling to master smuggled spinning jennies and carding machines. Slater landed in New York in November and went

151

straight to work for a local factory whose crude methods and lack of water power tried his patience.

Moses Brown of Pawtucket, Rhode Island, had exactly what Slater needed: the magnificent falls of the Blackstone River and a botched collection of Arkwright machines. Brown, who sprang from a dynasty of Providence merchants, had made his fortune and retired from trade at thirty-five. Now he was backing a cotton mill for his son-in-law William Almy and nephew Smith Brown. Slater wrote him, offering to perfect the machines. Brown replied, "If thou wilt come and do it, thou shall have all the profits made from them ... and have the credit as well as the advantage of perfecting the first water-mill in America." Brown's Quaker beneficence did not, in fact, impel him to give Slater all the profits, but history has given Slater full credit for what was really an inspired collaboration.

By January Slater was in Pawtucket. According to Moses Brown, "When Samuel saw the old machines he shook his head, and said, 'These will not do.' " They were housed in a rented fulling mill on the site of Joseph Jenks' forge. For ten weeks Slater worked with Brown's mechanics rebuilding a spinning frame. Having demonstrated his ability, he was taken into partnership and spent the rest of the year designing and helping to build two more spinning frames and the drawing, roving, and carding machines without which the frames were useless.

Almy, Brown and Slater went into production on December 20, 1790, and within a few days increased their staff from four to nine – seven boys and two girls, aged seven to twelve. Slater knew that the power of the mill depended not on the hands inside but on the water wheel outside. He later blamed the rheumatism that plagued him through life on the hours he spent early on winter mornings, breaking ice from the frozen wheel. Two years later when he, Almy, and Moses Brown's son Obadiah built their first mill, he covered the undershot water wheel with a shed.

Slater dominated the partnership because he knew what he wanted: specialization and continuous production. Gradually he persuaded Almy and Brown to concentrate on spinning yarn, giving up or cutting down work not yet adapted to water power such as weaving or knitting. What he could not make them understand was that the new power source demanded an operation as ceaseless as the flow of the Blackstone. They were used to the bespoke system, which spun yarn to fill customers' orders like a grist mill grinding farmers' grain. Slater's aim was to run at capacity, producing as much yarn as possible for sale on the open market.

Inevitably he became master of his own mill. In 1798 he built a much bigger factory across the river, advertised his yarn, and developed markets along the eastern seaboard. He kept his share in the Old Slater Mill, superintending both mills on a salary of $1.50 a day for each. In Samuel Slater and Company, his new and more amenable partners were his family: his father-in-law, Oziel Wilkinson, and his two brothers-in-law. (When he had arrived in Pawtucket Slater had boarded with Wilkinson, a blacksmith, and worked with him and his son David, a brilliant mechanic, building the Arkwright machines. The next year he had married Wilkinson's daughter Hannah.)

The heady air of democracy that suffused the new nation was an appropriate climate for systems that supplied goods in quantity to a population approaching four million, and Slater's success fostered competition. Alexander Hamilton's insistence that Congress should back cotton manufacture, the improvement of native cotton from the flimsy stuff Slater had once refused to spin, Eli Whitney's new gin and the importation of Samuel Crompton's mule, a cross between Hargreaves' jenny and Arkwright's frame, all sent merchants and mechanics scrambling into the cotton business. Men who had learned their trade from Slater spread his machines and methods through the States. His brother-in-law Smith Wilkinson, the child who ran the first carding machine in the Old Slater Mill, wrote, "I believe nearly all the cotton factories in this country, from 1791 to 1805, were built under the direction of men who had learned the art of building machinery in Mr. Slater's employ."

Among them were his relatives and partners. Smith Wilkinson ran mills in Pomfret, Connecticut, owned by his family who bought a thousand-acre site for the double purpose of keeping out taverns and providing farms for the parents of child workers. Almy and Brown bought the old Job Greene property at Warwick, Rhode Island, where a single building housed a grist mill, a sawmill, a trip hammer, a carding machine, and two tenements rented to families with remarkable powers of accommodation. More attractive to Almy and Brown was the cotton mill, whose millwright John Allen was dispatched to inspect the machines at the Old Slater Mill. He was taking measurements when Slater, furious that his improvements would be used in a mill in which he had no interest, threatened to hurl him through the window. Obadiah Brown, by now Slater's closest friend, gently grasped the yardstick: "I will finish thy work, and see if Samuel will serve me as he did thee."

Rivalry, like nepotism, was taken for granted in the labyrinthine alliances of early textile millers. Slater kept the link with his original partners, which had survived a much more serious dispute over the shared water rights of the two mills on the Blackstone. When his younger brother John finished his apprenticeship as a millwright, Slater brought him to America, took him into a partnership called Almy, Brown and Slaters, and installed him as manager of yet another cotton factory which grew to a mill village, Slatersville.

Presidents had paid fees to Oliver Evans; to Slater they paid homage. James Monroe visited his mill, Andrew Jackson visited his house. Though he had built up a fortune in mortgages, turnpike stock, and other investments, shortage of cash had forced him to sell his shares of the Old Slater Mill and Slatersville to William Almy in the depression of 1829. Crippled with rheumatism, he was plainspoken as always. Jackson called him the father of American manufactures, adding gracefully, "You set all these thousands of spindles at work, which I have been delighted in viewing, and which have made so many happy by a lucrative employment."

"Yes, sir," said Slater, "I suppose that I gave out the psalm, and they have been singing to the tune ever since." The vice-president remarked that he was glad Slater had also made money for himself. Slater answered, "So am I, for I should not like to be a pauper in this country, where they are put up at auction to the lowest bidder."

Strange comment, from a man who helped put in motion the machinery of management that drove North America into the era of industrial capitalism. Perhaps he sensed the contradiction at the heart of the Industrial Revolution: the vast new scale of power that freed workers to buy the products of their own labour would also enslave them to the manufacturers and the machines.

Archibald McArthur built a ▶ water-powered woollen mill on the Mississippi River at Carleton Place, Ontario, in 1871 and ten years later sold it to John Gillies. Bought by Bates and Innes in 1907, within two years it had 150 workers running knitting machines day and night.

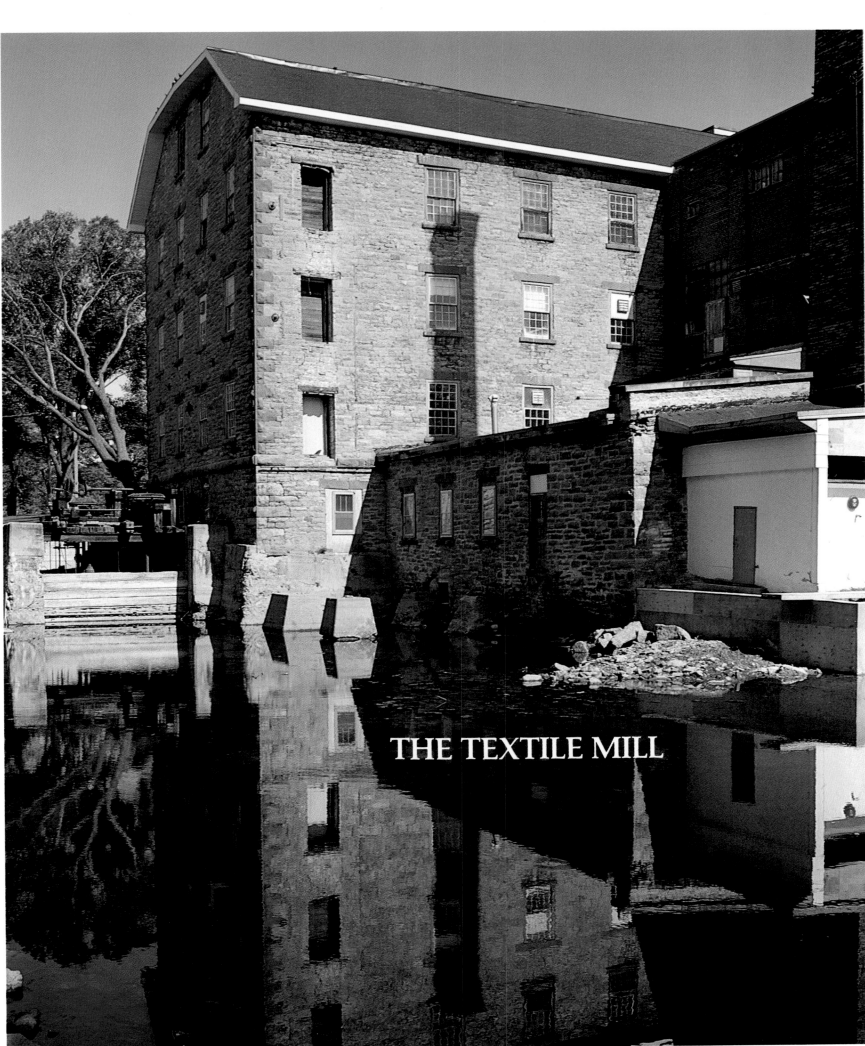

THE TEXTILE MILL

1

2

3

4

Pinnacles of Power

The belfries and cupolas above the great cotton mills of the northeastern states seemed to symbolize the soaring ambitions of their owners. After the Civil War the human scale and village customs of the early textile mills disappeared in the factory towns whose workers were faceless "hands" endlessly performing a monotonous task. When business slumped, employers cut wages and pressed for higher production, knowing that a slow or stubborn worker could be replaced from the swarm of immigrants begging for jobs. A petition for higher pay was circulated at Lowell; no one dared to sign it, and the three men who wrote it were traced, fired, and blacklisted throughout the industry. Past infancy, not yet mature, the American economy was in the throes of turbulent adolescence.

1. Looming above a substantial cornice, a tower at Harmony Mills is topped with ornamental ironwork.

2. The cupola with its weathervane surmounts the central tower of Pepperell Mills at Biddeford, Maine. Landed proprietors since the seventeenth century, the Pepperells invested some of their profits from lumbering and trade in this mid-nineteenth-century cotton mill whose products found a world market.

3. The heavy brick tower, ornately decorated without and hollow within, still serves as centrepiece for the enormous complex assembled from 1838 by the Amoskeag Manufacturing Company in the cotton-mill town of Manchester, New Hampshire.

4. The central tower of Ponemah Mills in Taftville, Connecticut. Taftville was a one-company town from 1871 – when Edward P. Taft from Providence, Rhode Island, opened his first mill – until the 1960's when the mills were sold.

5. The belfry of a cotton spinning mill built in 1814 by Eleuthère Irénée Du Pont beside his powderworks on the Brandywine near Wilmington, Delaware. Converted in 1884 to manufacturing metal powder kegs, the building now houses the Hagley Museum of industrial history.

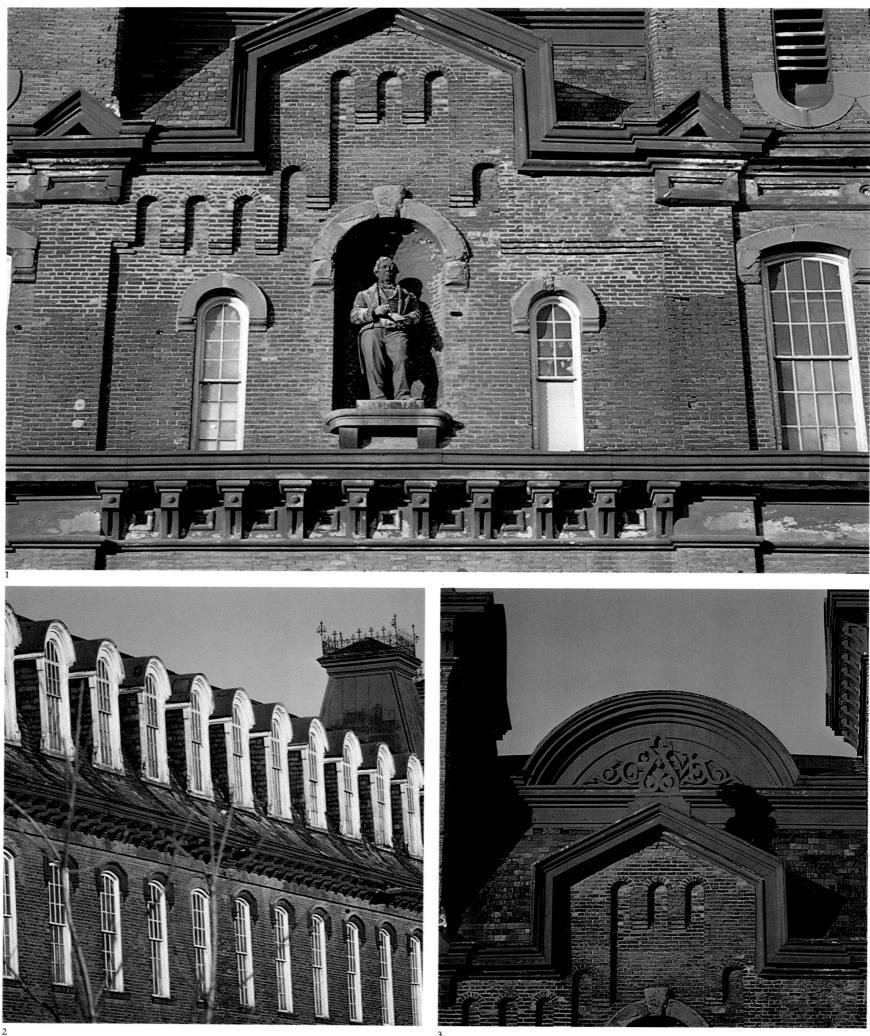

1

2   3

1. Peter Harmony stands enshrined above Mill No. 3 of the Harmony Manufacturing Company at Cohoes, New York. Begun in 1867 and completed in 1872, this monolithic edifice with its ecclesiastical overtones claimed to be the biggest complete cotton mill in the United States. Harmony was its first president and leader of the business group that formed the company in 1836. Continually expanding until 1872, Harmony Mills stayed in production until the Depression of the 1930's. Now the vast complex is occupied by several small industries.

2. Under the mansard roof with its handsome dormers, Harmony mill hands worked a minimum of ten hours every day except Sunday, for about $12 a week.

3. The twin towers of Mill No. 3 bracket the opulent stone and brickwork above Peter Harmony's niche.

4. James Rosamond's woollen mills in Almonte, Ontario, are now owned by Zephyr Textiles. Rosamond, an Irish immigrant, built his first mill in Carleton Place and moved to Almonte in 1857.

5. Ponamah Mills, built on a 600-acre site on the Shetucket River. In 1950, the mills were still running on water turbines supplemented by steam and electricity, producing thirty-five million yards of cloth a year.

6. The Amoskeag cotton mill stretches along the Merrimack River in Manchester, New Hampshire.

7. In 1881 the Trent Valley Woollen Manufacturing Company built this imposing textile mill at Campbellford, Ontario. The brick piers and corbelling strengthen the walls.

◀ *The first houses built for workers at Harmony Mills were small cottages. In the 1850's, the mill began packing its employees into rows of terraced housing east of Mohawk Street. Ten years later, as production expanded, row housing spread out on Harmony Hill. With more than a thousand workers, the complex became a self-sufficient company town within the city of Cohoes.*

Overleaf: Steam had replaced water power by the time this factory whistle was put on top of Nova Scotia Textiles in Windsor, Nova Scotia. Like the bell of a nineteenth-century cloth mill, its shrill sound governs the lives of the workers.

162

7

MEN WITH
NEW POWER

7

ENERGY AND EXPLOITATION

The nineteenth century was one of enormous contrasts. As it began, milling in the cities of the east was passing into the hands of businessmen with the capital to expand industries. On the remote shores of the Red River and Lake Superior, the Hudson's Bay Company and the rival North West Company were building mills to serve their furtrading posts. In the frontier states and Canadian provinces, pioneers were hacking out homesteads.

Like water seeking its level, the dispossessed of Europe flowed west to the New World. In Upper Canada, Loyalists were joined by restless Yankees, destitute Scots and Irish, and English hand craftsmen thrown out of work by the Industrial Revolution. The immigrants clustered in villages founded by millers: Merrickville, Streetsville, Spencerville, Wiltsetown, Cambridge Mills, Purdy's Mills (Lindsay), Scott's Mills (Peterborough), Wardsville (Smith's Falls) and hundreds of others. In Prince Edward Island in 1803, Lord Selkirk began his settlement at Belfast by building a mill on the Pinette River where John MacPherson still saws wood by water power today, and the grandsons of Samuel Leard of Tryon spread through the Island where now they run mills from Coleman to Cardigan.

From the first grist, saw, carding, and fulling mills sophisticated complexes developed. The wool and cotton factories begun in the 1830's grew into textile industries that made fortunes for the mill owners of Quebec and enabled Maritime families like the Stanfields of Truro and the Humphreys of Moncton to enter federal politics.

Throughout the century, the cotton industry served as the index of United States economy. Wool manufacture there was already declining by 1846, when James Rosamond built Upper Canada's first woollen weaving mill at Carleton Place. Fifty years earlier two Yorkshire immigrants, Arthur and John Scholfield, had opened the first water-powered woollen mill in the States at Byfield, Massachusetts, but they sold it in 1799, beaten by foreign competition. Five years later, when Arthur began making broadcloth with wool from Spanish merino sheep brought in to

◀ The Herr Mill near Strasburg, Pennsylvania, built as a grist mill about 1740, is now a gift shop and museum of early American firearms.

In Tryon, Prince Edward Island, Charles Stanfield from Yorkshire set up a farm, shipyard, hat factory, general store, and the woollen mill that inspired his sons to develop the world's first unshrinkable underwear.

improve the American strain, he convinced merchants he could equal imported quality, and James Madison's inaugural suit was tailored from his cloth. But while other wool makers prospered, the Scholfields wandered from state to state, setting up and selling out, and Arthur went bankrupt in the postwar depression of 1816 that ruined hundreds of other textile manufacturers.

Two things saved the cotton industry: the protective tariff of 1816 and the power loom, which extended mass production from spinning to weaving. In 1813, Francis Cabot Lowell drew on his memories of a tour of Lancashire mills to design a power loom for his Boston Manufacturing Company at Waltham, Massachusetts, the first mill in the world to convert raw cotton to finished cloth under one roof. Two years later William Gilmour, who had worked power looms in Scotland, offered to build one at Slatersville. John Slater said yes but Samuel said no, and Gilmour took his loom to Judge Daniel Lyman of Providence. With the help of a drunken English hand weaver who had come to watch and was instantly hired, Gilmour got his loom running in the Lyman Cotton Manufacturing Company. Lowell's machine cost nearly $300 and was held under patent. Gilmour's simpler machine could be built for $70, and Lyman offered the drawings to Slater's brother-in-law David Wilkinson, the first independent manufacturer of textile machinery, who typically added his own improvements. Even small mills could afford a loom that produced cloth for less than a tenth of a hand weaver's wage. In the first six months of 1826, four thousand were installed in Philadelphia alone, and a machinist who made seventy a week could not keep up with demand.

In 1820 the Boston Manufacturing Company began its immense mill complex at Lowell, Massachusetts, prototype for the textile towns first praised for production, later condemned for exploitation. The Pawtucket strike of 1824, protesting increased hours and reduced piece rates, was an inkling of unrest. For two weeks all mills but one were picketed and forced to shut down.

For the first time, women struck too. More than a hundred weavers, mostly girls in their late teens, met to argue the mill owners' contention that paying $2 a week plus board was morally wrong because it elevated women above their proper station.

The only romantic element in textile milling was the tycoons' vision of model communities where spirituous liquors were banned, mill hands were encouraged to save and invest, and children got religious and rudimentary education on Sundays. In 1827 Smith Wilkinson wrote from Pomfret that a seventy-two-hour week left no time "to spend in idleness or vicious amusement." In his twenty-nine years as manager, only two cases of seduction and bastardy had occurred. He allowed that, "Children under ten years are generally unprofitable at any price," but employed a few to help their parents. In overheated, underlit, unpartitioned mills filled with intricate, fast-moving equipment, workers were deafened by the cacophony, prey to respiratory disease from inhalation of fibres, and barefooted so the floor could be sloshed with water to humidify fragile cotton. Children dodged between machines, brushing rollers with hand cards and slipping inside moving frames to mend broken threads.

As the architecture of mill towns grew more elaborate, working conditions deteriorated. The Senate Committee of 1883 was shocked by testimony from mill hands dressed in rags, living on bread and clams, losing jobs to men whose children could work as back-boys for 30¢ a day. In Cohoes, New York, every building belonged to the Harmony Company; if water power for the mills ran short, boarding houses and tenements had no drinking water. A doctor from the textile town of Fall River, Massachusetts, reported that workers were dwarfed, undernourished, so exhausted that they drank night and day, and marked by a "dejected, tired, worn-out, discouraged appearance." A union leader added, "It is dreadful to see those girls, stripped almost to the skin, wearing only a kind of loose wrapper, and running like a racehorse from the beginning to the end of the day." The most important result of the committee was the establishment of a federal Bureau of Labor Statistics which provided a basis for bargaining between the industrialists and the rising unions.

TWO CANADIAN MILLS

More gradually, but just as irrevocably, other businesses, such as flour and sawmilling, grew too big for individuals, and were successful only if run by corporations. Nowhere was the difference between old ways and new more marked than in Quebec. Two flour mills in Lower Canada illustrate the contrast: le Moulin de

Gros-Sault, built by the Sulpicians in 1798, and the mill built by Alexander Ogilvie at Quebec in 1801. The fortunes of Gros-Sault rose and fell with the Canadien economy. Demolished in 1892, even its ruins have vanished from Ile Perry, and the rapids of Sault-au-Recollet that once drove it were flooded in 1928 by a power dam. Ogilvie's sons, taking all Canada as their province, founded a giant industry.

A typical seigneurial mill of the period, Gros-Sault was rectangular, fifty feet high, with stone walls five feet thick. Its tall roof had three dormers on each side, and the roof ladders, still seen on some country mills, that the miller climbed to throw down water if a fire started. One end had a double chimney, the other a wooden extension for carding and fulling machines. It cost 106,979 livres, a high price for the time, and the Sulpicians leased it to a series of millers.

In 1837 it was sold to Charles Perry, a Montreal tobacco merchant whose son got an excellent return of nearly $2,000 a year in the 1860's. But in 1876 other ventures put him into bankruptcy, and the mill was claimed by a trust company as mortgage security. Thus it came into the hands of Philias and Jean-Baptiste Prévost, lifetime residents of Sault-au-Recollet, who rented, ran, and eventually bought it for $16,000. In 1891 Jean-Baptiste sold it to a real estate agent who turned it over to the Montreal Water and Power Company for $80,000. One hopes that the Prévosts asked a good price since it seems fitting that the profit from the sale of a parish mill should benefit not an agent but the local man who operated it.

Le Moulin de Gros-Sault.

At Quebec the young Scots immigrant Alexander Ogilvie saw his market in the newcomers and lumbermen swarming into the city. He brought with him a pair of Seine valley burrstones and these, with a true-edge to test their level, a mill bill, a bar scale, and a slate for tallying incoming grain and outgoing grist, added up to an investment of perhaps $500. His mill at Jacques Cartier on the St. Lawrence shore had an undershot wheel which was connected by the hammered oak wedges of the pit wheel which in turn powered the stones on the second floor, which ground forty 196-pound barrels of flour a day.

Ten years later he moved to Montreal to join his parents, shipping his stones upriver to his uncle John Watson's mill at Recollet. In 1817 he married Watson's daughter, and took over the mill when her father died two years later. When he retired to his farm at Côte St. Michel, he passed on his stones to his brother-in-law James Goudie. Soon Goudie would get a much more valuable inheritance, the services of Ogilvie's sons.

With space for four to six run of stones and a capacity of a hundred barrels a day, Goudie's new mill on the Lachine Canal was the biggest in the country. When a mill next door threatened its water supply in 1851, young Alexander Walker Ogilvie persuaded his uncle to build the Glenora mill, four storeys high, with a steam engine to supplement the water wheel. Goudie retired in 1855, turning over A.W. Ogilvie and Company to Alexander, then only twenty-six, and his twenty-three-year-old brother John. Their younger brother, William, became the third partner in 1860.

Their success was archetypal. William managed milling and marketing in Montreal. Alexander mastered the finances of a notoriously volatile industry. Earlier in the century, the Corn Laws had given Canadian wheat and flour preference in Britain; by 1843, Americans were shipping grain to Upper Canada to be milled and exported as colonial produce. With the repeal of the Corn Laws in 1846, the price of Ogilvie Superfine flour dropped almost as low as grain. After 1854, when the Crimean War cut off Britain's supply from the Continent and the Reciprocity Treaty stimulated trade with the United States, business soared. Meanwhile John ranged Ontario and the prairies to find the best wheat, to build mills at Seaforth, Goderich, and Winnipeg, and, in 1881, the first grain elevator in western Canada at Gretna, Manitoba. It was he who saw that the future lay in the west.

OLDE GRIST MILL

RESTAURANT & COUNTRY STORE

RESTAURAN HOURS
WEEKDAYS
Luncheon 12:00
Dinner 5:30
SUNDAY
12:00-2:30, 5:30
Air Conditio
COCKTAI
COUNTRY S
& GIFT SHO
OPEN 11:00 A.M.
CLOSED MON

OPEN FOR THE SEASON

Left: Some mills that once ground flour now serve full meals with cocktails. At left, Perkin's eighteenth-century tide mill in Kennebunkport, Maine.

Above: A woollen mill in Goderich, Ontario, is now part of Benmiller Inn, whose owners use the turbines in their grist mill to heat the swimming pool.

Below: The Thompson-Perkins Mill at Richmond Landing in Ottawa has been bought by the National Capital Commission and transformed into a restaurant.

NEW SOLUTIONS TO OLD PROBLEMS

Again the United States had set the pace. By 1830, when the Baltimore and Ohio laid the first public railroad, a thirteen-mile track from Baltimore to Ellicott's Mills, the Erie Canal and the wheatlands of Ohio and upstate New York had already shifted the grain trade to Buffalo. Forty years later, thanks to the railways and the water power of the Falls of St. Anthony, Minneapolis was wheat capital of the world. Here modern flour milling was created by the first major improvements since those of Evans: the New Process, which used mechanical purifiers to produce vast quantities of white flour, and the transition from millstones to cylindrical metal rollers.

France and Hungary had long practised "high milling," a succession of grindings and boltings that yielded a range of grades from "white" for aristocrats to "black" for peasants. Democratic North America simply wanted a way to get the most nearly-white flour as possible from a bushel of wheat. They found it in the New Process, which combined fanning and sifting to purify middlings. Developed in France, it was introduced to Minnesota by the French engineers Nicholas and Edmund La Croix. In 1865 they set up purifiers for mills on the Cannon River built by immigrants from Canada, Alexander Faribault and the Archibald brothers, and, six years later, for the Washburn B mill of Governor Cadwallader Colden Washburn's Minneapolis Mill Company. The New Process increased the amount of saleable flour produced from a bushel of wheat so much that it gave Washburn a profit of $2 a barrel within three years.

Washburn and other Minneapolis milling magnates who quickly adopted purifiers had not yet grasped the dangers of dust inhalation and explosion from multiple bolting. "Miller's cough" had long been a symptom of lung disease. Large quantities of flour dust are highly explosive. The New Process required many more bolting machines. The Washburn A mill had 148 huge reels; bolting and fanning stirred up 3,000 pounds of dust a day. Since no one then understood that concentrating the dust made it even more likely to ignite, it was trapped in two rooms under the millstones for removal at night. On the evening of May 2, 1878, the Washburn A mill blew up, killing eighteen men and totally destroying the building and five other mills. Next morning Washburn visited the site of his new Washburn C mill, paced out an extension to its foundation and ordered a dust exhaust system.

In the new mill Washburn used not only millstones but chilled iron rollers, invented in sixteenth-century Italy for flattening metal and developed for flour milling in Switzerland by Jakob Sulzberg who installed them in Budapest's Walzmühle in 1839.

◀ *Set in a metal casing firmly mounted on a heavy floor are the flour rollers installed by Wilson Morningstar in his mill at DeCew Falls, Ontario. The rollers' high speed and uniform yield of fine flour made them ideally suitable for the big mill complexes of Minneapolis and other midwestern cities. As rollers came into use, millstones were discarded and many of the small mills that held them closed their doors forever.*

Tricky, expensive, still experimental, wasteful of bran and wheat germ that would sixty years later be replaced with vitamins and minerals, Hungarian rollers set the pattern for mass production. Washburn had thirty-six pairs cast in Connecticut for his A mill in 1874. The next year the first imported from Vienna were used in E.W.B. Snider's mill at St. Jacob's, Ontario. Within their lifetime the country millers would see their fine grindstones relegated to use as decorative doorsteps.

Though Minneapolis competition forced the Ogilvies to close all four mills briefly in 1887, they went right on building a new one. Using rollers Alexander had examined in Budapest, their Royal mill in Montreal ground 2,100 barrels a day, almost half as much as Pillsbury A, the biggest mill in the world. John's faith in Manitoba hard spring wheat had been vindicated. Two years earlier, when he sent to Scotland a small shipment of flour from his Winnipeg mill, the first exported from western Canada, the British Army and Navy tested it and astounded Ogilvie's by requesting half a million dollars' worth. The Ogilvies couldn't possibly fill the order, but now they knew that the market was there they refused to relinquish it.

Before he died in 1902, Alexander Ogilvie saw cereal replace timber as Canada's leading product. His enterprise was yeast for the industry, and his career rose with it. In 1884, as a senator, he was escorted by William Van Horne to the Canadian Pacific railhead at Medicine Hat. According to legend, Van Horne quipped, "The national flower of Canada? Ogilvie flour, of course."

In Upper Canada, millers survived similar cycles of ruin and reward. The War of 1812 filled ports with hungry soldiers while it drew farmers off their land into the militia. Flour doubled in price and distilleries were closed – for a while – to conserve grain for milling. In November 1814, American invaders burned mills through the southwestern counties, and the Backhouse Mill at Port Rowan was one of only a few left standing. Two companies of the 104th Regiment from New Brunswick were stationed to keep the mill at Ball's Falls grinding for troops. Billa La Rue's mill near Mallorytown, also under guard, ran nights and Sundays.

By the 1830's, Lower Canada was deep in depression. Soil exhaustion, crop failures, insect plagues, floods, droughts, and cholera carried by immigrant ships brought the habitants close to starvation. The harvest of 1836 was the worst on record. Upper Canada had prospered on grain, timber, and road and canal building, but both provinces were now demanding a true voice in

"Don't give up the ship!" was the legendary command of James Lawrence in the memorable battle between the British Shannon and American Chesapeake in 1813. In fact the American frigate surrendered; after serving in the British Navy until 1820, she was broken up and her timbers were used as floor beams for the Chesapeake watermill at Wickham, Hampshire, in England.

government and their unrest spawned the Papineau and Mackenzie rebellions of 1837.

Again millers joined the militia; John Backhouse's mill was run by his children, aged thirteen and fifteen. Other millers, suspected of opposing the Family Compact, had to flee villages they had founded. William Purdy of Lindsay was jailed without trial in Cobourg; released, he moved to Bath. Daniel Shipman of Shipman's Falls, now Almonte, watched from his mill window as a cavalry troop rode up to arrest him. He escaped and hid out for months until the authorities saw the futility of trying to convict the town's first citizen.

Roused by William Lyon Mackenzie, patriots in the United States revived the old dream of freeing Canada from British rule. In November 1838, the secret Hunters' Lodges organized an attack on Fort Wellington at Prescott led by Nils Von Schoultz, a former Polish officer, and Bill Johnston, the renegade of the Thousand Islands. Johnston, more successful as guerrilla than as naval commander, bungled his part, leaving Von Schoultz and two hundred invaders to take refuge in the eighty-foot stone tower mill at Windmill Point. For five days they held it, under siege by regulars and militia until the arrival of three steamers carrying 24-pounder guns forced surrender. Though the raiders were defended at their court martial by a promising young King-

In the Battle of the Windmill, invaders who had hoped to capture a fort ended by surrendering a tower mill. Here, in 1838, two hundred men from the United States crossed the St. Lawrence in a vain attempt to take Fort Wellington in the Canadian town of Prescott. Armed opposition forced them to retreat to this stone mill. Five days later they surrendered, and eleven were sentenced to death.

ston lawyer, John A. Macdonald, Von Schoultz and ten others were hanged, sixty men were exiled to Van Diemen's Land, and the rest were released to report that Canada was not ready for liberation.

The 1850's were the heyday of milling in Ontario. Wheat crops that trebled in the decade, free trade with the Maritimes, reciprocity with the States, the Erie Canal route, and the raging railway boom excited financiers. Wheat, the only saleable crop, became a cash crop when the millers decided to form their own bank – The Bank of Toronto.

Toronto already had a dozen banks, many shaky, but rural districts had almost none. Farmers preferred barter to currency, which circulated in wild and sometimes worthless varieties. What cash they had they kept under their mattresses, an invitation to housebreakers who roamed the province in 1855. Millers were becoming exporters, and so much money changed hands that they hoped to keep profits from loans, brokerage, and insurance inside the grain trade. Led by William Gamble, who owned a five-storey flour mill, a woollen factory, and a lumber business, millers from the Toronto-Hamilton area won a charter, rounded up share capital and opened for business at 78 Church Street, Toronto, on July 8, 1856. Millers, merchants, farmers, and lumbermen streamed in to deposit and borrow.

The first two presidents were bankers; the third, William Gooderham, knew milling as well as money. When he was appointed in 1864, his nephew James G. Worts was already vice-president. Now famous as distillers, Gooderham and Worts had earlier tried a much more quixotic enterprise: a seventy-one-foot windmill on the Don River. Worts' father had come from England in 1831, leaving thirteen-year-old James in Montreal while he looked round Upper Canada. Deciding to settle, he sent for James with instructions to bring not only their luggage but machinery for a windmill. Sixteen days later the boy reached York, now Toronto, in a hired bateau crammed with machinery and a crew of six Indians. Twelve builders began working, and though some were often "drunk as David's sow" the mill was ready when the main shaft and millstones arrived next spring. Worts then asked his brother-in-law William Gooderham to come to York with both their families. He landed that fall with forty-two Gooderhams and Wortses plus eleven small orphans adopted when their parents died aboard ship. The windmill sails blew off in a gale three years later, but the tower lasted as a landmark for the Old Windmill Line for many years.

This oil painting by an unknown artist ▶ *shows Gooderham and Worts' York Windmill at the mouth of the Don River in Toronto.*

FROM TUBWHEELS TO TURBINES

Most windmills were already gone. As late as 1855 a Rochester company proposed building fifty to grind flour on the prairies, but the characteristic windmills of the west would be fan-bladed irrigation machines and backyard generators. Now power would come from a direct descendant of the old Norse wheel – the turbine.

Some of the first New England millers had used tubwheels: horizontal water wheels in round wooden housings. A ten-foot waterfall could turn wooden paddles set in a vertical wood shaft directly attached to the millstone above. The wheels were simple, slow, and inefficient, but they seldom iced up. Early in the nineteenth century Americans began patenting improved versions of submerged horizontal wheels in which the tub filled with water to put extra pressure on the runner vanes. These reaction wheels, using the casing as a working part of the machine, doubled the power of the old tubwheel.

The first person (except, predictably, Leonardo da Vinci) to invent a true turbine was Benoit Fourneyron of France. His design of 1827 introduced a stationary inner cylinder with a second set of blades feeding water out to a circle of rotating vanes curved in the opposite direction. The machine was housed in an iron casing that could be submerged to run under ice. Entirely immersed, it pulled in all the available water and used it to run continuously, like the earlier reaction wheels and unlike Poncelet's vertical wheel which made full use of water only as it entered and flowed out. Because the heavy fixed guide blades and light runner vanes worked against each other, the direction of water flow reversed within the motor, creating twice the pressure of a reaction wheel. The impulse of water flowing smoothly into the centre of the machine produced energy; the reaction of water flowing out doubled the amount of energy, like the recoil of a gun. Fourneyron's first turbine used 87 per cent of the force of the water, a much higher rate of efficiency than any previous water wheel.

Other engineers quickly began designing different types of turbines, and improvements continued to increase their efficiency. In 1844 Uriah Boyden installed the first American turbine in the cotton mill town of Lowell in Massachusetts. Soon James Leffel and other manufacturers were mass producing turbines to replace old-style water wheels in mills of all kinds. A miller who switched to a turbine might have to rechannel his water supply, but once the new system was working it ran all winter, increased his production, and freed him from the continual mending demanded by a wheel susceptible to weather damage.

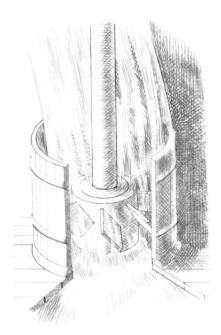

Early tubwheels were simply horizontal water wheels housed in wooden barrels to protect them from icing up in winter. A Norse wheel turned with the current of the stream; tubwheels were usually set below falls that directed water to the blades at the base.

Opposite: As electricity came into use ▶ late in the nineteenth century, many small mills began lighting up their neighbours' houses. At Perth, Ontario, citizens decided to spend town funds on rebuilding the old Tay View grist mill, mending the dam and installing a Leffel turbine and equipment to turn the mill into a hydro-generating plant that supplied the town from 1896 to 1922. Anson Bowes took over the idle mill in 1929 and ground flour till the mill burned in 1952. Now he has restored it as a museum, which still produces electricity for his house and farm.

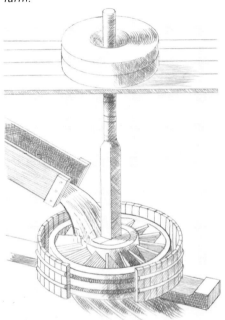

The reaction wheel was a transitional stage between tubwheel and turbine. Tubwheels were driven only by the impulse of water striking their blades. Reaction wheels were designed to use this force twice, first as it hit the blades and again as it rebounded back to them from the walls of the casing. In the early 1800's Americans worked out a prodigious number of variations on this concept. Here a flume delivers water to the slanted blades of a wheel housed in an inner wooden casing, shown in cutaway. Only the outer tub is stationary. The inner case, wheel, and shaft revolve to turn the upper millstone.

TIDES, DAMS, AND SLUICES

1. At Perkin's Mill in Kennebunkport, Maine, the control dam and gate at left allow the incoming tide to flood the millpond. As the tide ebbs the gate is closed.

2. After nearly eighty years of grinding, the Peirce Mill on Rock Creek near Washington, D.C., stopped abruptly in 1897. Alcibiades D. White was grinding a load of rye for a neighbour when the main shaft broke. The customer hauled away his unground grain, and the mill stood idle until it was repaired as a WPA project in the Depression and restored in the more prosperous sixties as a display of pioneer industry.

3. In 1936 fourteen citizens of Weston, Vermont, founded the Vermont Guild of Old Time Crafts and Industries. They bought a local sawmill with a 150-year-old dam and restored its turbines. In 1938 a flood washed out the dam and undermined the mill. Next year they rebuilt the foundation and began collecting mill machinery. Now the Old Grist Mill, part of a restored village, holds an unselective but absorbing collection of artifacts from both grist and sawmills.

4. The grist mill built for John Winthrop, Jr., at New London, Connecticut. Son of the governor of Massachusetts, he set up the first ironworks in both states. By 1847 he was governor of Connecticut. His enquiring mind ranged the sciences, and he was the first colonist elected to the Royal Society after it was founded in 1662.

1

2

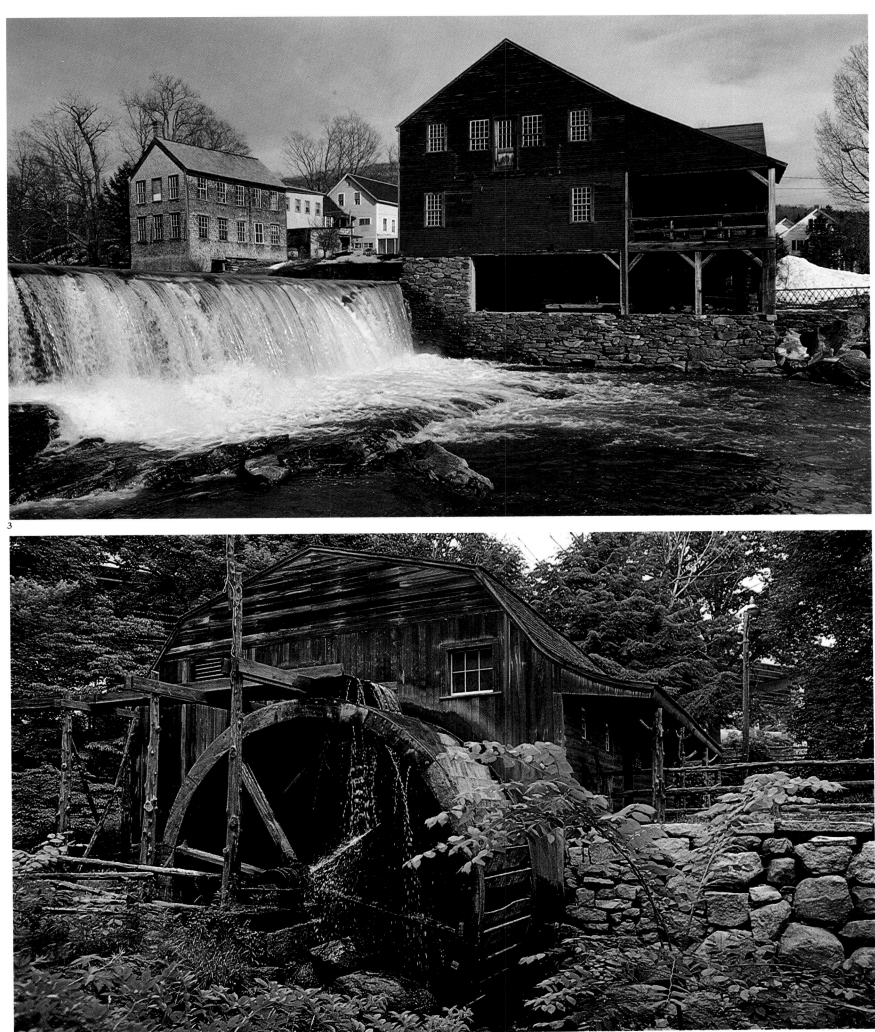

3

4

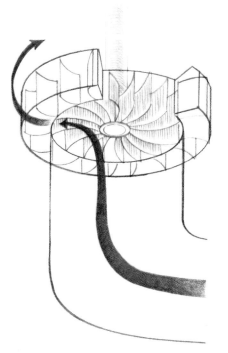

By the outbreak of the Civil War, New England and the central states were using turbines in many of the mills that made the North industrially self-sufficient. The South had neglected manufacturing for cotton growing. In the city of Lowell alone there were more cotton spindles than in all the mills in the eleven states that seceded. The North imported cotton from abroad and looted it in the South. Paper mills began using wood pulp, and flour millers sold their smaller orders in paper bags instead of cotton sacks. Woollen factories were revived to run day-round and week-long to make uniforms and blankets for the Union Army. Weapon works such as Colt at Hartford, Connecticut, and other factories that were quickly switched to war tools were so productive that New England could afford to go on exporting her surplus arms to Europe.

The Union Army had plenty of gunpowder from the five factories of the Du Pont family in Delaware and Pennsylvania. Their machinery was driven by three steam engines and forty-seven water wheels, most of them turbines. On their 2,000-acre site on the Brandywine River, the firm also ran three woollen mills, a cotton mill, and a flour mill. From the Eleutherian Mills set up near Wilmington in 1802 by Eleuthère Irénée Du Pont de Nemours, who had studied under the French master chemist Lavoisier and was appalled by the poor quality of American powder, the firm had expanded to become the biggest powderworks in the United States.

The foundries, forges, and powder mills of the South were emergency operations, vulnerable to explosions and northern raiders. All factories were short of skilled machinists and metal workers, and even ordinary hands were conscripted for the Confederate Army and replaced by slaves, women, and children. "Nearly all the lathes are idle for want of hands," wrote the commander of the Naval Iron Works at Atlanta, adding that he was "unable to have forged the wrought iron bolts for the Brooke gun for the want of blacksmiths."

The Union Army laid waste Georgia, South Carolina, and Virginia, sacking and burning mills, barns, and houses. From Virginia, where seventy mills and two thousand barns stored with grain had been destroyed in a single expedition, General Lee wrote in January 1865, "There is nothing within reach of this army to be impressed; the country is swept clean.... We have only two days' rations." The wrecked plantation economy proved no match for the diversified, industrialized northern states.

Fourneyron's turbine could be fed from above or below. Here, water is delivered by a flume from below. This requires either a high fall of water, or a millpond whose level is at least equal to that of the turbine. The turbine's fixed central guide blades curve the water flow as it passes out to the rotating runner vanes. Slanted in the opposite direction, the outer vanes twist the water like a snake as it flows through them and out to the tailrace. The shaft, which is attached to the outer runner vanes, runs up from the turbine to drive the mill machinery.

By the end of the century, America was looking for a workhorse turbine, cheaper and less temperamental than Fourneyron's. In 1890 James Leffel developed his "Samson" model. The water flowed into the turbine through the narrow gates (as the arrow indicates), changed direction as it hit the rotating inner vanes (shown in cutaway), and again as it flowed out through the "buckets" at the bottom. By varying the pitch of the vanes, the miller could adjust the water flow to suit his location, water supply, and purpose.

THE WEALTH OF THE FORESTS

In the same decade, sawmilling replaced the trade in ships' timber throughout the United States and Canada. New Brunswick had always been a lumberman's landscape, its white pine forest dotted with logging camps and roads, its rivers clogging with bark and edgings, its ports crowded with timber ships. Otty and Crookshank built the first steam sawmill at Saint John in 1822, only a year after steam was first used to cut wood in Maine. In the 1830's the shrill whine of circular saws for lath and edging sounded alongside the thud of up-and-down saws; twenty years later, they were whirling in the main gates of sawmills. On the Miramichi River, Gilmour and Rankin's Newcastle mill was the biggest in the province until Joseph Cunard built a steam-powered mill with five gangs of reciprocating saws and three circular saws at Chatham in 1836. That was the boom year in which scores of mills sprang up along the Saint John River. Many were financed by Americans who went broke in next year's depression, which hit even Cunard. In the 1840's the most popular site became the river mouth, since it was cheaper to float down logs than sawn lumber which arrived sodden and stained. By 1850, New Brunswick had 4,300 men working in about six hundred sawmills, most of them small water-powered outfits on streams. Within a generation many would be abandoned as the industry concentrated in the complexes of the Saint John and the Miramichi rivers.

Amos Peck Seaman, the most ebullient Maritimes magnate of the nineteenth century, claimed that he built the first steam sawmill in Nova Scotia. On August 13, 1843, he rejoiced in his journal , "On Thursday last, 10th inst., steam mill first set in motion so the 10th day of August may be called the anniversary of the Minudie steam mill, the first ever in this province. I have built this mill at an expense of fifteen thousand pounds.... Bring in your logs, thirty thousand or more, She will saw them up before the year is O'ver." Three years later he built Grindstone Palace in Minudie, a mansion embellished with the finest mahogany fetched by his own ships and flagstones from his own quarries. Once penniless, barefoot, unschooled, he made a fortune from his Boston-centred interests in shipping, mining, property, and the biggest grindstone trade in North America. After he was presented at court in England by Lord Derby he proclaimed his social rise by rechristening himself – King Seaman.

In the Canadas, lumber on the St. Lawrence River, the Trent River, and Georgian Bay never rivalled the Ottawa Valley, that thriving, brawling, messy empire of wood. Ottawa, chosen by Queen Victoria in 1857 as the capital city of Canada, began as lumber country where fifteen years earlier the traveller John Godley had found "stumps scattered through the gardens of the houses, and pine trees through the streets." As late as 1884 the future prime minister, Wilfrid Laurier, admitted, "I would not wish to say anything disparaging of the capital – but it is hard to say anything good of it."

Philomon Wright.

Philomon Wright, who founded Wrightsville, the first settlement in the district, would have disagreed. In 1799 he surveyed his site, now the centre of Hull, Quebec, from "the tops of one hundred or more tall trees" and found his future power in the Chaudière Falls. Next February he left Woburn, Massachusetts, with five families, twenty-five axemen, livestock, tools, and provisions. By 1803 he had a sawmill and a grist mill. He floated his first raft of logs down to Quebec in 1806, just when Britain needed Canadian timber to replace the Baltic supply cut off by Napoleon's Berlin Decree.

Wright was an unstoppable organizer and a hopeless manager. By 1826, when Bytown, later renamed Ottawa after the Indian tribe of the region, was founded across the river, he had gathered a community of a thousand, built the Columbian Hotel and the first road in the region, and started the first passenger boat service on the Ottawa River. He was also so deep in debt to "Damned Horse Leeches" that his affairs were handed over to trustees. His son wrote, "I am out of money. How we shall get on God only noes. Don't hire any more Irishmen."

With the building of the Rideau Canal, Bytown eclipsed Wrightsville. Its completion opened the south shore to shipping and unleashed a flood of navvies into settlement and lumbering. Jean-Baptiste Saint-Louis, who had worked on the canal, built the town's first small water-driven sawmill at the Rideau Falls in 1830. The mill's third owner, Thomas McKay, had founded New Edinburgh and put the profits from his masonry contracts for canal locks and the Union Bridge into milling. In 1833 he built a sawmill, the first merchant flour mill on the Bytown side and, eight years later, a woollen mill where he installed power looms in 1847. By 1850 he had a world market, a gold medal from the London Exposition for his blankets, a seat in the Legislature of Upper Canada and 1,100 acres of bushland in Rockcliffe. Here he built his house Rideau Hall, now the governor general's residence.

Today the prime minister lives in the house once owned by a federal member of Parliament, Joseph Currier, who built the Long Island Mill on the Rideau at Manotick, Ontario, in partner-

A painting of Wright's mills at Chaudière Falls.

ship with the town's founder Moss Dickenson. For the opening of the mill in 1860, Campbell's Hotel catered a cold collation for sixty guests. Next March, Currier, just married, took his bride to see the mill. In what *The Ottawa Citizen* called a "Melancholy and Fatal Accident," Mrs. Currier was warned to be careful descending the stairs, "but scarcely were the words spoken when the unfortunate lady ceased to exist. A portion of her dress came in contact with the shaft and she was torn round with a frightful velocity and crushed against a pillar."

THE NEW WORLD IN A NEW AGE

By the 1860's the timber trade of New Brunswick and Ontario had yielded to the sawn lumber industry, dominated by entrepreneurs like the Cunards, the Gilmours, the Gillies, Perley and Pattee, Ezra Butler Eddy, Henry F. Bronson, and John Rudolphus Booth. The Reciprocity Treaty of 1854 and the States' demand for boards, planks, and deals launched the era of the American lumber barons.

In 1851, E.B. Eddy from Vermont rented a shack from the Wrights where he and his wife made matches by hand, peddling them by horse and buggy. In 1870 he built his first sawmill. In 1882, when he was mayor of Hull and a member of the Quebec Legislature, his mill burned. From Quebec City he wired, "Put out fire. Clean up debris. Prepare to build. Will be home tomorrow." The same spirit carried him through the fire of 1900 which destroyed all his plants along with the entire centre of Hull and most of the mills and factories at the Chaudière in Ottawa. Canada's lumber age was ending. Eddy and Booth, who led production for a generation, had switched to making pulp when Booth sent the last ceremonial timber raft down the Ottawa in 1904.

In the United States, settlement had pushed lumbering to the frontier. In 1850, Canada and the United States each had about 50,000 sawmill hands, but that year's American production of nearly $60,000,000 topped even Canada's peak years at the end of the century. By 1890, when Michigan led the lumber trade, almost 150,000 hands worked in 25,708 American mills. Ceaselessly, sawmilling, like textile and flour manufacture, moved toward consolidation, higher profits, and western expansion.

The greatest of contrasts lay between the century's opening and its close. Spreading across the continent, the United States had forsaken the agrarian past for the industrial future. Canada's dominion had been bound into nationhood by rails of steel. The mills of North America had been transformed by new inventions, by new methods, and by the new breed of corporation millers.

With sluicegates open, the ▶ *spring runoff comes surging over the dam beside the Hope Sawmill near Keene, Ontario.*

HARNESSING
WATER POWER

1

2

Channelling the Flow from Pond to Mill

For centuries millers have been working out ways of getting more power from water. First they dammed streams to create a reservoir and increase the head of water striking the wheel. Flow from the dam was controlled by sluicegates and channelled through a flume to the water wheel. Vertical wheels, vulnerable to ice, were housed inside or replaced by turbines that operated below the surface.

From wheel or turbine, power was transmitted through a chain of gears to grindstones, saws, tilt-hammers, or textile machinery. The working parts of an old mill, with hand-hewn cogs and rough wooden flumes, look simple but represent the first practical use of many engineering principles.

1. MacLaren's Mill at Wakefield, Quebec, is set below a natural waterfall given added power by a cement dam, built about 1906 to replace an outworn jumble of planks and boards. The grating below the bridge filters flotsam from the water before it enters the turbine. The cart above the dam controls the water level by adding or removing logs in the sluicegate.

2. A long raceway carries water to the wooden overshot wheel of the Backhouse Mill north of Port Rowan, Ontario.

3. Well-worn handles adjust the valve that controls the flow of water to the turbine at Grein's Feed Mill in Mount Forest, Ontario, built by James Dodds in 1866.

4. The control cart moves along the top of the dam at Watson's Long Island Mill at Manotick, Ontario.

5. The flume carries water to the turbine at Brant's Mill in Ligonier, Pennsylvania.

3

4

5

Water Wheels & Turbines

1. This fourteen-foot water wheel once powered the Old Red Mill in Clinton, New Jersey, now a museum.

2. Fed by a wooden sluice from the dam, the grist mill at Brewster, Massachusetts, grinds cornmeal for summer visitors.

3. At the Saddle Rock Tidal Grist Mill in Nassau County, New York, the control gate that regulates the speed of the water wheel is operated by the spoked wheel in the foreground.

4. The twenty-six-foot overshot wheel of McClung's Grist Mill in Zenith, West Virginia, develops about eighteen horse-power when the stream runs high.

5. The twenty-foot oak water wheel was added when the grist mill at Colvin Run, Virginia, was renovated in 1970. Built before 1820, the mill has a canal and sluiceway almost 2,400 feet long.

6. To avoid icing in winter, the water wheel of the Legendre Mill at Stornaway, Quebec, is inside the building. Built by Télesphore Legendre in 1883, it ground flour until about 1940.

7. The wheel at Roblin's Mill in Black Creek Pioneer Village stands idle, sheathed in ice.

8. After its water wheel was replaced by two turbines, the grist mill at Delta, Ontario, ran summer and winter until 1949.

9. Blades at the base of the turbine power the reciprocating saw of O'Hara's Mill near Madoc, Ontario.

7

8

9

1

1. At Grein's Mill in Mount Forest, Ontario, gears to lift the sluicegate are connected to the penstock control that determines the flow to the turbine.

2. A lantern pinion wallower at Roblin's Mill transmits power from pit wheel to main shaft.

3. Cogs on the wheel at right turn the spur wheel that drives the hopper boy at Roblin's Mill.

4. The shaft from the turbine below carried power to the main bevelled gear wheel at Krug's Sawmill in Chesley, Ontario.

5. Powered by the turbine at Grein's Mill, the main bevel gear meshes with the small gear that drives the main shaft.

6. The belt running from this wheel in the cellar of Ancaster Mountain Mills, Ontario, controls equipment upstairs.

7. A drive belt is hand-spliced with leather lacing at the Ancaster grist mill. In the heyday of the region, four hundred men worked in sixteen mills powered by a common source, the Fountains in the Hills.

2

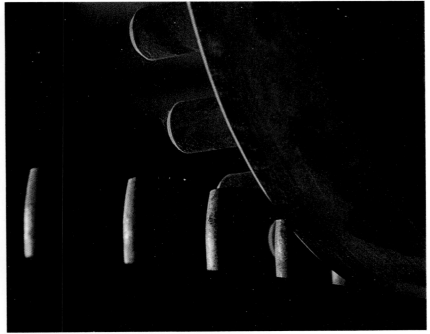

3

4

5

6

7

At MacPherson's Grist Mill in Pictou County, Nova Scotia, the shaft of the smaller wooden gear in the foreground runs up through the bed stone to drive the runner stone on the floor above. This gear is one of four planetary gears around the circumference of the great spur wheel, centre right. Two of the other planetary gears also run millstones; the fourth, a drive gear, turns the great spur wheel. This drive gear is in turn driven by shafts, belts, and wheels, that connect with the mill turbine in Sutherland's River. More than a hundred years old, the magnificent wooden machinery of the mill is being renovated by Dr. Howard Locke of New Glasgow.

8

VOICES OF
THE PAST,
VISIONS OF
THE FUTURE

Helen Fox

8

LIVING WITH MEMORIES

As milling passed into the hands of corporations with capital, research teams, and huge market facilities, the small mills began gradually dropping out of business. New power sources – steam, then electricity – enabled companies to set up complexes in ports, in major cities, and in the regions that supplied their raw materials. Railways and highways left the old mills marooned far from trade routes. The great forests of the east were gone, and lumbermen moved west or opened factories to make pulp for newsprint from spruce and second-growth woodlands. The introduction of new building materials such as steel and concrete meant that less lumber was needed. The textile industries concentrated and eventually developed synthetic materials.

Through the first half of this century, the old mills gave up one by one. The communities that once depended on them turned to the network of suppliers that spanned the continent. Donald Robson, a descendant of the Morningstar family who ran mills at Font Hill, Ontario, from 1883 until 1940, is now curator of the family mill which is restored as a museum. He says, "Technology caught up with these mills. The business was running down over a number of years. There were a lot of mills in this area, and it was pretty hard to compete. The big mills kept taking more and more of the business, and for the small mills it was slam, slam, slam – and out."

Mills owned by families for generations were abandoned as sons and daughters moved off the land into city professions. Millers enlisted in the two world wars, and some never returned. No one was left to mend wooden buildings, dams washed out by spring and fall floods, or wornout machinery.

In New Brunswick, Cecil Jeffries turned down a partnership in his great-grandfather's mill at Sussex Corner after his father's cousin sold it. "In a mill that age the machinery was very old, very heavy, and when anything went wrong it cost a small fortune to get it repaired. The new owner couldn't get the hang of dressing stones, and the farmers stopped bringing in buckwheat. You could only work the sawmill in the spring when the water

◀ The Spring Mill constructed in 1817 in Mitchell, Indiana, is one of hundreds of local mills that flourished in the nineteenth century until they were bypassed by the railway system thrusting west.

was high. Even the business of selling ice from the pond was going out as people got refrigerators. I didn't see how I could make a living at it, and I don't think I made a mistake."

Most of the old millers faced the same problem: they couldn't make a living. They were proud men, reluctant to be forced out of business, eager to stay in by moving to a similar trade such as bulk feed, willing to help transform their mills into museums. Hundreds of mills powered by wind and water still exist, tucked in almost forgotten corners of our states and provinces. Some have been restored as showcases for the science and tradition of the past. More are ruins, tumbled heaps of stone or wooden sheds with buckled roofs overhead and dusty trails of sawdust underfoot. Others are still producing stone-ground flour, animal feed, fine woollens, and lumber, and the men and women who work or once worked in these living mills can tell us about them.

Raymond Dunnell of Bernardston, Massachusetts, is still called in by the mill he once owned to splice rope. It is needed for an extraordinary power transmission system, one of those modifications rigged up by an ingenious miller. When his grist and sawmill on the Falls River burned out in 1895, Charles Barber built a five-storey feed mill beside highway and rail tracks, driven by a turbine in the river attached to a continuous cable of 1,450 feet of one-and-an-eighth-inch manilla rope that runs through the mill wall at the second storey and down to the main shaft in the basement to power the whole mill. Between river and mill the rope passes through a twenty-four foot tower that supports it and houses an idler to take up slack when it stretches in wet weather. The idea of running a big mill by a kind of high-speed ski tow is pure Rube Goldberg, but it's been working for eighty years without a hitch.

Now the Bernardston Grain Company does more bulk trading than grinding, but in the fifties J.L. Dunnell and Son did an annual $250,000 business in custom work for farmers. Dunnell says, "I was twenty years old when we bought the mill in 1921, and my father and I worked together as long as he was alive. He'd been in the business before I was born and he enjoyed it – and so did I. We did everything from bookkeeping down to greasing the machinery. The old money drawer was so worn that we had to patch the bottom of the change cups. That drawer never had a lock on it, and only once or twice we had a small amount of money stolen. We've left thousands of dollars' worth of feed on that porch overnight and never lost more than a few bags."

Around the grist mill that was built by ▶ Andrew and David Schriver in 1797 grew the village of Union Mills near Westminster, Maryland. The Civil War split family loyalties. The Schrivers in the original homestead sent two sons to the Union Army; four of their cousins joined the Confederates.

On June 30, 1863, General Jeb Stuart's Confederate cavalry swept through the town, seizing food, horses, and a night's rest. As they left, hard on their heels came the Union Army under General James Barnes. Within twenty-four hours, ten thousand soldiers were quartered at Union Mills. Next day the two great armies met at another village – Gettysburg, where the defeat of Robert E. Lee turned the tide against the South.

Helen Fox

Guy S. Kister, the oldest millwright in Ohio when he died in 1975, was eighty-three when in 1967 he built a new overshot wheel for his grist and sawmill, the only Ohio mill still powered by a wooden wheel. No one else knew the art his family had learned through three generations of making wheels for themselves and others. He chose the best white oaks from his own woodlot, felled and hauled them to his sawmill, and designed and installed an eighteen-and-a-half-foot wheel.

Herman Masse of East Vasselboro, Maine, is just as reluctant to give up water power. He is building an overshot wheel on a shaft and hub from an old New Hampshire steamboat, though he plans it only as a supplement for the diesel engine that now drives Masse Lumber Company. It was the last big sawmill in New England to run on water power till August 1972, when the shaft from the biggest turbine suddenly snapped. "So we switched to diesel," Masse says, "but we've got a new shaft and if there's a fuel shortage or the diesel breaks down, we can switch back to water power."

Masse has turned the mill over to his son Kenneth but he still works there. He inherited his mechanical aptitude from his father Louis, a Quebeçois from Trois Rivières who bought the mill in 1912. "I never worked for a pay cheque but for three months in my whole life," says Masse. "After I graduated from college my father said, 'We can go in business together or you can go it alone.' I decided to go it alone, and I took over the sawmill in 1926. In the fifties things began to change so fast that it got to a point where you'd either get in business or get out. We decided to fabricate what we needed ourselves and it's worked out pretty well, with a little Yankee ingenuity. I never made a blueprint for anything. I just go ahead and build it. I've got a picture up here in my head."

Flora Abbott of South Waterford, Maine, grew up in her father's mill too. When the Hapgood Carding Mill was moved to Old Sturbridge Village, Massachusetts, for meticulous restoration, she wrote, "I must confess that the tears came to my eyes when the first truck came to start hauling it off, as I remembered how, when as a little girl I started to help, I had to stand on a stool to put the wool on the batting machine and how father insisted that we must pug up our hair before he would allow us to go among the cards; at the time I thought it rather a hardship, but with a larger understanding with the years, I knew his always wanting some of his children with him whatever he was doing, and always finding someway they could help was a good part of our early training and made good workers of us all."

The ramshackle ramp leads to a ▶ derelict grist mill in South Melville, Prince Edward Island.

RAVAGED
BY TIME

201

Relics and Ruins

Deserted mills have a gentle, elegiac quality quite different from the mood of a working or restored mill. Doorways are overgrown with a tangle of wild raspberries and young birches; ridgepoles sag like a swaybacked horse; field flowers thrust new life through gaps in the tilted floor boards.

Windmills stand impotent, their vulnerable sails wrecked by storms. Wooden water wheels are rotted by long immersion or dried, cracked, and warped by exposure to sun and ice. Mill races choke with driftwood, ponds fill up with marshland, and dams are reduced to a rubble of rocks and broken cribwork.

Some mills seem to have been abandoned as suddenly as the mystery ship *Mary Celeste*. If your fancy leads you down Mill Street or along the old river road, you may find a sawmill left empty for years. Yet planks and sawdust still litter the floor, a rotary saw rests on supports riddled with rust, and water flows through the penstock to the idle turbine underneath the mill. Often the back of the building sits crazily astride the sloping riverbank, supported by stilts planted in the river.

In derelict mills, time hangs suspended, heavy with history. A sense of the miller's presence is evoked by stair treads worn by his feet, leather belting spliced by his hands, timbers notched and pegged by the dusty tools on his workbench. As each mill subsides into oblivion, a fragment of the past is lost forever.

Once a handsome building, completely sheathed in shingles, the Moors Sawmill in Wilton, New Hampshire, is slowly rotting out.

2

3

1. Except for the turbine cover in the foreground, the burned-out foundation of the Harris Woollen Mill at Rockwood, Ontario, looks more like a Greek ruin than a mill.

2. Rusting out at Sharparoon, New York, this metal water wheel still shows the welding scars left by buckets long since vanished.

3. At North Tryon, Prince Edward Island, the sawmill once owned by George Ives has a buckled roof and toppled chimney.

4. Green tendrils of ivy twine round machinery discarded from a mill at Port Elgin, New Brunswick.

4

1

2

From Innovation to Obsolescence

These ruined mills and their machinery were pacesetters in their day. Turbines were a nineteenth-century innovation, considered vastly superior to old-fashioned water wheels. Today, all but a few have been replaced by diesel engines or been left to rust away in abandoned mills. The hydro mill at Cataract was producing electricity before the great generating stations were built at Niagara Falls.

1. Water spills busily past the turbine housing of the ruined sawmill at Breadalbane, Prince Edward Island.

2. This ravaged foundation once supported one of the earliest water-generated electric systems in Ontario. From the falls at Cataract in the Credit Valley, John Deagle began supplying electricity to the village of Erin in 1899, for a fee of one cent per bulb per night. By 1909 the Cataract Electric Company was serving Erin, Alton, and Orangeville. The pioneer plant was eventually sold to Ontario Hydro and phased out in 1947.

3. Beside Darr's shingle mill in North Wentworth, Nova Scotia, lies a discarded turbine, its ten large curved blades housed in wooden planking bound with iron bands and fixed to a wooden centre shaft running to an iron gear wheel fitted with wooden cogs.

Overleaf: Its years of grinding past, an old millstone is put to new use propping a timber under John Hagerman's grist mill near Woodstock, New Brunswick. The tiny building was the Hagermans' first home when they married during the Depression.

GOOD TIMES AND BAD

Millers are matter-of-fact people, yet their memories of good times and bad are suffused with a sense of work well done and regret for a way of life that disappeared with their generation. Jack Leard of Crapaud, Prince Edward Island, sold his mill in 1953 because he had no son to carry on after him. He remembers, "When my people owned the mill it was a flour mill, sawmill, shingle mill, carding mill, and electric light plant, and we cut ice from the millpond in winter. My father and my grandfather before me owned it, and the mill went back a hundred years before that. I grew up in the mill – I started in when I was seven or eight years old. The Island was a great place for mills. The farmers would come in for news and gossip and they'd be there all day. A lot of community life centred around those old mills."

Fire could turn a carding mill into a torch in minutes. The only one left in New Brunswick is Briggs and Little Woolen Mill at York Mills, owned by Ward and Roy Little. Ward Little has worked there sixty years, and he remembers how the mill burned out in 1944. "The fire started in the feed of the cards. We think it must have been a match in the wool. I noticed smoke coming up the side of the cards and went up to get a hose, but she went too fast. Belts started dropping through the floor, glass was breaking, smoke was rolling out the windows. I never even had time to shut the mill off."

One of the first portable steam sawmills in Nova Scotia destroyed a building even before it was put inside. In the early 1890's Alexander Sutherland of Denmark in Colchester County ordered it from an Ontario manufacturer who brought it down by train and helped set it up. Sutherland's son Wilfred says, "Father put up a shed to house it while he built the permanent building. Well, that portable mill burned in the night and it took the new building with it. There was only a short stack on them portable steam mills – the cinders and everything flew out from the furnace. Father's big new rubber drive belts, that he'd just got that morning, were right inside the door and he stood there and watched them melt away. He even lost his overalls. There was no insurance in those days, and he had to start over again and build the mill here that I was brought up in. Nearly every young fellow round this part of the country earned their first dollar in there in the thirties."

In Nova Scotia, portable mills set the life-style of the lumber camps of the early twentieth century. Russell Hayman of Drysdale Falls learned to play a violin before he could handle a saw. "Usually somebody had a fiddle or maybe a banjo," he says, "and we used to sing. It was hard work but it was something most everybody liked, working in the woods. It was a healthy job and you got good food, bread and beans and potatoes and meat and black tea. I was in partnership with my three older brothers, and I started when I was fourteen shovelling sawdust out from under the saw. Then I went to burning slabs off the logs; nobody'd think slabs was worth anything in them days. Then I took away wood from the big saw after it came off the carriage; then I was edging man; then I went to canting, turning the log; then to sawing. Sawyer was the top job. You had to split up the log, gauge how many planks could be cut. Mostly all the logging was done in the winter time and you just set up the sawmill in the woods. Generally the first thing you done was to build a camp – camps in them days had an upstairs for the men to sleep – and then you set up the sawmill on notched logs. You'd work till the lumber was gone and then you'd move maybe twenty miles. The mill and the camp was all taken apart, every stick, and moved with wagons and sleds on those rough roads through the woods. The boiler weighed five ton and it took eight horses to haul it."

"It was a good life." Again and again the phrase recurs. Even Laura Mullins, who spent the First World War working in Stanfield's first knitting mill in Tryon, Prince Edward Island, revived by new owners to make longjohns for troops, says, "We girls liked it great. The machines were so noisy you'd never hear the water wheel, and when you came in you'd wonder if you could ever hold your head straight to get out, but you got used to it. My father worked twenty-nine years sanding doors in Mr. George Ives' sawmill at North Tryon. It ran off water from the old Stanfield dam. He loved it, but he had to stop because he got ulcers and the doctor blamed the sanding machine. He drank buttermilk for two months after that, and he lived to 103."

THE POWER OF THE FUTURE

Today we take for granted the hope of longer life and leisure bequeathed to us by the millers of the past. The industrial age that destroyed most of the old mills had its roots in the technology evolved in mills through centuries. A good miller was always searching for ideas to increase his profits, expand his business – and make his work easier. Exploring new methods, adapting new machines for their own purposes, using new power sources, millers exercised their minds as well as their muscles to push industry forward. The systems of automation and mass production first developed in milling have utterly changed the way we live.

MILLS PRESERVED AND RESTORED

1. The old mills of Virginia have special value because few were built while early farmers concentrated on tobacco, and many were destroyed in the Civil War. The Colvin Run Mill in Fairfax County, built before 1820, ground more than a million bushels over a century. In 1969 it was reconstructed with prematurely aged materials.

2. The Wight Grist Mill on the Quinebaug River, built of vertical planking on a stone foundation, was one of the earliest buildings reconstructed at Old Sturbridge Village.

3. As the War of Independence ended, Colonel Nathaniel Burwell and General Dan Morgan of the Continental Army put prisoners of war from Hessian regiments to work building their mill in Millwood, Virginia. Their millwright, L. H. Mongrul, was a resourceful fellow; challenged to a duel, he avoided the encounter by choosing as weapons sledge-hammers wielded in six feet of water.

4. In this composite drawing of the Gilbert Stuart Birthplace in Saunderstown, Rhode Island, the small mill in the foreground was built in 1757 to grind cornmeal. Historically more important is the house on the right, where the painter Gilbert Stuart was born in 1755. His father, a snuff grinder from Glasgow, had been brought to America by Dr. Thomas Moffat of Newport, who hoped to compete with merchants who imported the fashionable mixture of tobacco and spices from England. The snuff mill inside Stuart's house was an edge-runner mortar-and-pestle machine driven from an outside water wheel through a series of wooden gears.

In 1765 Moffat was collector of taxes imposed by the detested Stamp Act. He fled to England but is still hanged in effigy once a year in Rhode Island. Ironically, Gilbert Stuart grew up to paint the portrait of Washington used on stamps and dollar bills.

1

2

3

4

We are in debt to the millers who have cut our work week to a fraction of the time they laboured.

Hydro-electricity, first generated as a by-product of the turbines of watermills, provided power to light their villages. To the old Norse wheel we owe the generating stations of watershed areas such as the Great Lakes. Their towers striding through slashed forests and their cables slung along country roads remind us that the generators are descendants of the watermills they have rendered almost obsolete. Many parts of North America and other continents must depend on electricity produced by burning fossil fuels.

As oil and natural gas sources dwindle and nuclear energy is still loaded with unknown hazards, scientists are re-exploring the clean inexhaustible powers of wind and water. Throughout the world, dozens of projects are testing giant windmill installations to produce electricity. Like early windmills, they face the key problem of how to store energy from a fluctuating source for use when they need it. Though no one has yet found how to make large-scale wind generators mechanically and financially feasible some scientists predict that they could supply up to 20 per cent of the United States' electric needs by the year 2000. At the other end of the scale, small fan-bladed machines are again being used for houses and cottages where power lines are unavailable or extravagant. No system is perfect, but many are working.

Like wind, ocean water is free, non-polluting, available – and costly to harness on a large scale. But experts on energy are testing ways of drawing electricity from the tides of the Minas Basin in Nova Scotia and the River Rance in Brittany, the steady current of the Gulf Stream, and the waves of the Pacific and Atlantic. Optimistic researchers believe that floating wave systems pumping water into turbogenerators could provide enough power to meet the present electrical requirements of the entire United Kingdom. What all these ambitious schemes have in common is the miller's instinctive faith in natural power.

While scientists search for new ways of using natural power, others are beginning to recognize the historical value of old wind and water mills. Since its first meeting in 1965 the International Symposium on Molinology has been collating the research of professional millwrights around the world. In North America, the Association for the Preservation of Technology and the Society for the Preservation of Old Mills keep members posted on academic studies and grassroots gossip. National, provincial, state, and local organizations are protecting mills that remain, and many individuals have restored old mills to working order or have converted them to houses, restaurants, shops, and craft centres. The range in quality is wide. Some local museums house an indiscriminate jumble of unrelated mill artifacts. Others have been planned with such scrupulous attention to detail that they sweep the visitor back into the period of the mill's heyday.

Only recently has technology been put to full use in North American renovations. Some aesthetically pleasing restorations do not stand up to close inspection of their workings. A grist mill rebuilt in the 1930's has its grindstones a floor's-length away from the face wheel, a waste of power transmission that would horrify a miller. One of the prettiest watermills in New England grinds with massive stones from a much bigger mill. Ironically, the sails and water wheels of many restored mills now turn idly while conventional electricity drives their grindstones. But the enthusiasm of historical societies, combined with the expertise of museum researchers and scientists, promises well for the mills that remain.

New concern for nutrition and plain good eating has brought some old grist mills back to life. Whether stone-ground flour with all its natural food values really is healthier than flour enriched with chemicals still provokes argument, but it certainly *tastes* better. After eating bread baked from buckwheat flour water-ground by MacPhersons Mills in Nova Scotia, sampling smoky oats freshly kiln-dried over a slow fire of burr-rock maple at Balmoral Mills, and baking Rhode Island johnnycake with white cornmeal from Prescott Farm or Kenyon's Grist Mill in Usquepaugh, you wait a while before you buy your next loaf from a store.

Nothing except actually being there can convey the feeling of health and harmony inside a working mill. Looking up at the sweep of great sails, listening to water splashing through a wheel, and feeling the surge of irregular rhythms as the wooden cogs and rumbling stones carry on their ancient craft convey a sense of well-being. There is satisfaction in touching new-spun yarn, watching a huge log inching its way through a thudding sawmill, or fingering warm meal spilling from grindstones treasured through generations.

Active or abandoned, the old mills stand as testimony to the diligence and ingenuity of millers. They tell a story of harnessing natural power, of patient modifications and flashes of genius, and above all of community life. Magnificent, shabby, deceptively bucolic, they offer a glimpse of the energy of our ancestors and new hope for the survival of our children.

1

2

WARP, WEFT, AND SHUTTLE

MacAusland's Mills

Here, on a tributary of the Mill River, Archibald MacAusland built a water-powered sawmill with an up-and-down reciprocating saw, and a flour mill that ground wheat with burrstones brought from France. His son Fred added a carding machine to process wool for local sheep farmers. In 1932 he installed equipment for custom spinning and weaving wool.

On September 6, 1949, a high wind carried a flash fire in the picking machine through the mill, leaving buildings and their contents a mass of rubble. Fred MacAusland, then seventy-five, died two days later, his lifework destroyed. Determined to carry on, his sons Edward and Reginald searched the United States for machinery to rebuild. By New Year's Eve their new mill was in operation, and they still weave pure wool blankets from the fleece of sheep raised in the Atlantic Provinces.

1. South of Bloomfield, Prince Edward Island, the highway dips into a hollow where the brown-shingled buildings of MacAusland's Woolen Mills cluster beside the road. Behind the mill is the pond with its banks of red Island clay.

2. In the loft, fleeces wait to be spun and woven into blankets. The raw wool is first washed, then oiled to replace the natural lanolin. In the picking room, where wool fibres fly like snowflakes, it passes through the picker, a roller studded with tiny spikes that shake out any remaining dirt and fluff the wool before it goes to the carding machine.

3. A lap of soft wool is fed into the cylinders of the carding machine which combs it into a thick rope – the sliver – which is passed to the drawing frame. This is one of the few carding machines still in commercial use in North America.

1

2

Spools and Shuttles

1. Loose, irregular yarn is pulled from spools underneath the roller in the foreground. Each strand, fed through its own groove, is stretched and tightened as the roller moves back and forth to left and right while it rotates. The spools behind await their turn to be placed below the roller.

2. Here the spools are feeding yarn up to the grooved roller above.

3. From the roller at top, yarn is passed to the long reel which twists three strands together to feed three-ply wool to the bobbins below.

4. The flying shuttles of the loom move swiftly from side to side carrying the weft across the warp. All the odd-numbered threads of the warp are lifted while the weft passes under them. The even-numbered threads are lifted as the weft returns. The power loom was essentially a hand loom adapted to other power sources.

5. The shuttles with their sharp metal-tipped ends must be handled by a deft and experienced weaver.

3

4

5

The town of Parshallville, Michigan, ▶
grew round the flour mill built by
Isaac Parshall about 1869. It was
renamed Tom Walker's Grist Mill after
the Walker family who ran it for nearly
half a century. Until 1970 it was still
operating as a cider mill, not a grist
mill, and its present owner plans to
restore it to working condition.

BIBLIOGRAPHY

ANDREANO, RALPH (ed.). *The Economic Impact of the American Civil War*. Cambridge, Mass.: Schenkman Publishing Co., 1967.

ARMSTRONG, JOHN BORDEN. *Factory Under the Elms: A History of Harrisville, New Hampshire, 1774-1969*. Published for the Merrimack Textile Museum by the M.I.T. Press, Cambridge, Mass., and London, England.

BAGNALL, WILLIAM R. *The Textile Industries of the United States*. Cambridge, Mass.: The Riverside Press, 1893.

BARCK, OSCAR THEODORE, AND LEFLER, HUGH TALMAGE. *Colonial America*. New York: The Macmillan Company, 1968.

BATHE, GREVILLE AND DOROTHY. *Oliver Evans, a Chronicle of Early American Engineering*. Philadelphia: Historical Society of Pennsylvania, 1935.

BAYARD, FERDINAND-MARIE. *Travels of a Frenchman in Maryland and Virginia, 1791*. Translated and edited by Ben C. McCary. Ann Arbor, Mich.: printed by Edwards Bros., 1950.

BECK, J.H. *Liber Tertius de Ingeneis*. Milan: Edizione il Polifilo, 1969.

BENNETT, RICHARD AND ELTON, JOHN. *History of Corn Milling*. 4 vol. New York: Burt Franklin, 1964. Originally published London, 1898-1904.

BISHOP, J. LEANDER. *A History of American Manufactures from 1608 to 1860*. Philadelphia: Edward Young and Co., 1868.

BOYCE, GERALD E. *Historic Hastings*. Belleville, Ont.: Ontario Intelligencer Ltd., 1967

BRISSOT DE WARVILLE, JACQUES PIERRE. *New Travels in the United States of America, 1788*. Edited by Durand Echeverria. Cambridge, Mass.: The Belknap Press of Harvard University Press, 1964.

BROWN, GEORGE W. (gen. ed.). *Dictionary of Canadian Biography*. Toronto: University of Toronto Press, 1966.

BROWN, HOWARD MORTON. *Founded Upon a Rock: Carleton Place Recollections*. Carleton Place, Ont.: 150th Year Festival Committee, 1969.

BURSTALL, AUBREY F. *A History of Mechanical Engineering*. London: Faber and Faber, 1963.

BURT, A.L. *The Old Province of Quebec*. Toronto: McClelland and Stewart Ltd., Carleton Library, 1968.

CASEY, THOMAS W. "Napanee's First Mills and Their Builder," Ontario Historical Society Papers and Records. Toronto: Ontario Historical Society, 1905.

DEFEBAUGH, JAMES ELLIOTT. *History of the Lumber Industry of America*. 2 vol. Chicago: The American Lumberman, 1906.

DE LITTLE, R.J. *The Windmill, Yesterday and Today*. London, John Baker, 1972.

DEMOS, JOHN (ed.). *Remarkable Providences, 1600-1760*. New York: Braziller, 1972.

DOUVILLE, RAYMOND AND CASANOVA, JACQUES. *Daily Life in Early Canada*. New York: The Macmillan Company, 1967.

DUNHAM, MABEL, "Mills and Millers of Western Ontario," *The Northwestern Miller*, February 2, 1936.

EVANS, OLIVER. *The Young Mill-wright and Miller's Guide*. Philadelphia: printed for, and sold by the author, 1795.

FAIRBAIRN, WILLIAM, *Treatise on Mills and Millwork*. 2 vol. London: Longman, Green, Longman and Roberts, 1861.

FARRIES, K.G. AND MASON, M.T. *The Windmills of Surrey and Inner London*. London: Charles Skilton, Ltd., 1966.

FAUTEUX, JOSEPH NOËL. *Essai sur L'Industrie au Canada sous le Régime Français*. Quebec: Ls-A. Proulx, Imprimateur du Roi, 1927.

FINCH, WILLIAM COLES. *Watermills and Windmills*. London: The C. W. Daniel Company, 1933.

FREESE, STANLEY. *Windmills and Millwrighting*. Cambridge: Cambridge University Press, 1957.

GARBER, D.W. *Waterwheels and Millstones: A History of Ohio Gristmills and Milling*. Columbus: The Ohio Historical Society, 1970.

GARRATY, JOHN A. (ed.). *Labor and Capital in the Gilded Age*. Boston: Little, Brown, 1968.

GELDERS MOLENBOEK, n.v. Drukkerit en Uitgeverij de Walburg pers Zutphen, 1968.

GORDON, EVELYN, AND GRANT, HARRY. *The Vanished Village*. Petheric Press Limited, 1972.

GUILLET, EDWIN C. *Early Life in Upper Canada*. Toronto: The Ontario Publishing Company Limited, 1933.

HAIG, ROBERT BRUCE. *Ottawa, City of the Big Ears*. Ottawa: Haig and Haig Publishers, 1970.

THE HAGLEY MUSEUM. Wilmington, Delaware: Eleutherian Mills-Hagley Foundation, 1957.

HINDLE, BROOKE. *America's Wooden Age*. North Tarrytown, N.Y.: Sleepy Hollow Restorations, 1975.

HOPKINS, J. CASTELL. *Progress of Canada in the Nineteenth Century*. Toronto: The Progress of Canada Publishing Company, 1900.

HOWELL, CHARLES AND KELLER, ALAN. *The Mill at Philipsburg Manor*. North Tarrytown, N.Y.: Sleepy Hollow Restorations, 1975.

HUGHSON, JOHN W. AND BOND, COURTNEY C.J. *Hurling Down the Pine*. Old Chelsea, Que.: The Historical Society of the Gatineau, 1964.

HUTSLAR, DONALD A. "Ohio Waterpowered Sawmills," *Ohio History*, Winter-Spring 1975. International Symposium on Molinology. *Transactions of the Second Symposium*, Denmark, May, 1968. Brede, Lyngby, Denmark: Danske Mollers Venner, 1971.

JACKSON, A.T. *Mills of Yesteryear*. The University of Texas at El Paso: Texas Western Press, 1971.

JOHNSON, ALLEN (ed.). *Dictionary of American Biography*. New York: Charles Scribner's Sons, 1957.

JONES, ROBERT LESLIE. *History of Agriculture in Ontario, 1613-1880*. Toronto: University of Toronto Press, 1946.

KALM, PETER. *Peter Kalm's Travels in North America, 1750*. The English version of 1770, revised from the original Swedish and edited by Adolph B. Benson. New York: Dover Publications, Inc., 1966.

KIRKCONNELL, WATSON. *County of Victoria Centennial History*. Lindsay, Ont.: Victoria County Council, 1967.

LANCTOT, GUSTAVE. *Canada and the American Revolution, 1774-1783*. Toronto: Clarke, Irwin and Co., 1967.

LEGGET, ROBERT FERGUSON. *Ottawa Waterway*. Toronto: University of Toronto Press, 1975.

LESCARBOT, MARC. *History of New France*. Toronto: The Champlain Society, 1907-14.
Lore and Legend of Brome County. Knowlton, Que.: The Brome County Historical Society, 1965.

MALONE, DUMAS, AND RAUCH, BASIL. *American Origins to 1789*. New York: Appleton-Century-Crofts, 1960.

MATTICE, PAUL B. "Old Grist Mills Along the Schoharie," *County Historical Review*, Schoharie County Historical Society, May 1949.

McKENZIE, R. *Leeds and Grenville: their first two hundred years*. Toronto: McClelland and Stewart, 1967.

MEAD, DANIEL W. *Water Power Engineering*. New York: McGraw-Hill Book Co., 1915.
Missisquoi County Historical Society. *Eighth Historical Report*. 1965.

MUNRO, WILLIAM BENNETT, *The Seigneurs of Old Canada*. Toronto: University of Toronto Press, 1964.

PATTERSON, GEORGE. *A History of the County of Pictou, Nova Scotia*. Montreal: Dawson, 1877.

PIERSON, WILLIAM. "Harrisville, New Hampshire," *Antiques*, October 1972.

POTTER, MARGARETTA WOOD. "How We Saved the Jamestown Windmill," *Newport History*, Winter 1975.

PRÉVOST, ROBERT. *Le Moulin de Gros-Sault*. Montreal: Les Editions Archonte, 1939.

RAWLYK, GEORGE A. *Revolution Rejected, 1775-1776*. Scarborough, Ont.: Prentice-Hall of Canada Ltd., 1968.

REABURN, PAULINE. "Power from the Old Mill Streams," *Canadian Geographic Journal*, March, 1975.

REYNOLDS, JOHN. *Windmills and Watermills*. London: Hugh Evelyn, 1970.

ROGERS, COL. GEORGE D. "History of Flour Manufacture in Minnesota," St. Paul, Minn.: Minnesota Historical Society, 1905.

ROY, PIERRE-GEORGES. *Vieux Manoirs, Vielles Maisons.* Publié par la Commission des Monuments Historiques de la Province de Québec. Quebec: Ls,-A. Proulx, Imprimateur du Roi, 1927.

SCHULL, JOSEPH. *100 Years of Banking in Canada: A History of the Toronto-Dominion Bank.* Toronto: Copp Clark, 1958.

SKILTON, C.P. *British Windmills and Watermills.* London: Collins, 1947.

SMEATON, JOHN. *Experimental Enquiry Concerning the Natural Powers of Wind and Water to turn Mills and Other Machines.* London: printed for I. and J. Taylor, 1744.

STANLEY, GEORGE F.G. *Canada Invaded 1775-1776.* Toronto: A.M. Hakkert Ltd., 1973.

STEVENS, G.R. *Ogilvie in Canada: Pioneer Millers, 1801-1951.* Toronto: McLaren and Son Ltd., 1957.

STOKHUYZEN, FREDERICK. *The Dutch Windmill.* New York: Universe Books, Inc., 1963.

STORCK, JOHN, AND TEAGUE, WALTER DORWIN. *Flour For Man's Bread: A History of Milling.* Minneapolis: University of Minnesota Press, 1952.

STRUTHERS, E.J. "The Early Settlements of the Eastern Townships." Lecture to Extension Department of Bishop's University, 1972.

SYSON, LESLIE. *British Water-Mills.* London: B.T. Batsford Ltd., 1965.

THOMSON, ETHEL AND ARCHIE (ed.). *History of North Tryon, P.E.I.* North Tryon: North Tryon Women's Institute, 1973.

The Tread of Pioneers: Annals of Richmond County and Vicinity. Richmond, Que.: Richmond County Historical Society, 1966-68.

TUNIS, EDWIN. *Colonial Craftsmen.* Cleveland, Ohio, The World Publishing Co., 1965.

TUNIS, EDWIN. *Colonial Living.* Cleveland, Ohio: The World Publishing Co., 1957.

TUNIS, EDWIN. *Frontier Living.* Cleveland, Ohio: The World Publishing Co., 1961.

VAN EVERY, JANE. *With Faith, Ignorance and Delight: Homer Watson.* Aylesbury: Homer Watson Trust, 1967.

WAILES, REX. *Saxstead Green Mill, Framlingham, Suffolk.* London: Her Majesty's Stationery Office, 1960.

WAILES, REX. *Windmills in England: a study of their origin, development and future.* London: Architectural Press, 1948.

WALKER, HARRY J. *100 Years: Ottawa and the Valley.* Ottawa: reprinted from *The Ottawa Journal,* 1967.

WHITE, GEORGE S. *Memoir of Samuel Slater.* Philadelphia, 1836.

WRIGHT, LOUIS B. *The Cultural Life of the American Colonies, 1607-1783.* New York: Harper and Row, 1957.

WYNN, GRAEME CLIFFORD. "The Assault on the New Brunswick Forest, 1780-1850." Ph.D. thesis, University of Toronto Graduate Studies, 1974.

ZIMILES, MARTHA AND MURRAY. *Early American Mills.* New York: Clarkson N. Potter, Inc., 1973.

MANUSCRIPTS AND PERIODICALS

APT Bulletin, quarterly. The Association for the Preservation of Technology. Mrs. Ann A. Falkner, Secretary-Treasurer, 2030 Thorne Avenue, Ottawa, Ontario, K1H 5X5.

Heritage Canada, quarterly. Terry McDougall, Editor, P.O. Box 1358, Station B, Ottawa, Ontario, K1P 5R4.

Old Mill News, quarterly. The Society for the Preservation of Old Mills. Donald W. Martin, Editor, P.O. Box 435, Wiscasset, Maine 04578.

Public Archives of Canada, Chalmers Manuscripts.

Public Archives of Nova Scotia, James Barry Diaries and Agricultural Petitions.

ACKNOWLEDGEMENTS

The authors are especially grateful to the following people who shared with us their extensive regional knowledge of mills, or their experience as working millers:

Drummond Birks, Montreal, Quebec; Judy Boss, The Nova Scotia Museum, Halifax, Nova Scotia; Clifton Boyd, Middletown, Rhode Island; John Cullity, East Sandwich, Massachusetts; John O. Curtis, Old Sturbridge Village, Massachusetts; Derek Drummond, Montreal, Quebec; Raymond Dunnell, Bernardston, Massachusetts; Russell Hayman, Drysdale Falls, Nova Scotia; Catherine Hennessey, Heritage Foundation, Charlottetown, Prince Edward Island; Heritage Canada, Ottawa, Ontario; Charles Howell, Philipsburg Manor, North Tarrytown, New York; Gene Hughes, Frankfort, Kentucky; Cecil Jeffries, Sussex Corner, New Brunswick; George Kimball, Prescott Farm, Middletown, Rhode Island; Gary Kulik, Old Slater Mill Historic Site, Pawtucket, Rhode Island; Pierre Lahoud, Ministère des Affaires Culturelles, Gouvernement du Québec; John Leard, Crapaud, Prince Edward Island; Warren Leard, Coleman, Prince Edward Island; Ward Little, York Mills, New Brunswick; Edward MacAusland, Bloomfield, Prince Edward Island; Archibald MacDonald, Balmoral Mills, Nova Scotia; Donald W. Martin, Wiscasset, Maine; John L. Martin, The Nova Scotia Museum, Halifax, Nova Scotia; Herman Masse, East Vasselboro, Maine; John McAvity, The New Brunswick Museum, Saint John, New Brunswick; Mr. and Mrs. Jim McClure, Chesley, Ontario; Laura Mullins, Tryon, Prince Edward Island; Mary Peck, Historical Resources, Fredericton, New Brunswick; Robert Power, Historical Resources, Fredericton, New Brunswick; Everett S. Powers, Glenside, Pennsylvania; J. Rémillard, Institut d'Historie de l'Amérique Française, Montreal, Quebec; Donald W. Robson, Fonthill, Ontario; Wilfred Sutherland, Denmark, Nova Scotia.

We also want to thank federal, state, provincial, county and municipal historical societies, museums, conservation authorities, and individuals whose help and encouragement made our work easier. A special word of thanks is due to our research assistant, Nora Elliott.

It would be impossible to produce a book of this kind without drawing on the work of previous writers, and we are in debt to the authors listed in the bibliography.

PICTURE CREDITS

INDEX OF MILLS

This listing includes all existing mills that are illustrated in this book.

Acton Feed Mill, Acton, Ont. (n.d.), 62

Adam's Sawmill, Dewitt Corners, Ont. (c. 1828), 106, 109

Adamson's Sawmill, York Mills, N.B. (pre-1875), 112

Allan's Grist Mill, Dewitt Corners, Ont. (mid-1850's), 57

Amoskeag Textile Mills, Manchester, N.H. (c. 1838), 156, 159

Ancaster Mountain Mill, Ancaster, Ont. (c. 1863), 57, 193

Backhouse Mill, Port Rowan, Ont. (c. 1798), 57, 188

Ball's Grist Mill, Jordan, Ont. (c. 1809), 61, 127

Balmoral Grist Mill, Balmoral Mills, N.S. (c. 1873), 49, 84

Barrington Woolen Mill, Barrington, N.S. (c. 1884), 125

Benmiller Woolen and River Mills, Goderich, Ont. (1850's), 171

Bois de Boulogne Moulin à Vent, Paris (n.d.), 40

Bowes Grist/Hydro Mill, Perth, Ont. (c. 1823), 179

Brandywine Textile Mill, Wilmington, Del. (c. 1814), 157

Brant's Mill, Ligonier, Penn. (c. 1700's; reconstructed 1926), 84, 189

Breadalbane Sawmill, Breadalbane, P.E.I. (n.d.), 53, 206

Bridgehampton Windmill, Bridgehampton, Long Island, N.Y. (c. 1820), 24, 26

Burwell-Morgan Mill, Millwood, Va. (c. 1785), 103

Chatham Windmill, Chatham, Cape Cod, Mass. (c. 1797), 28

Cheshire Mill, Harrisville, N.H. (c. 1847), 133, 134

Colvin Run Grist Mill, Fairfax County, Va. (pre-1820), 210

Cornell Mill Museum, Stanbridge East, Que. (c. 1830), 83

Crane Paper Mill, Dalton, Mass. (c. 1844), 57

Darr's Shingle Mill, North Wentworth, N.S. (1890's), 108, 207

Deagle Hydro Mill, Cataract, Ont. (1850's; burned 1899, converted to generating electricity), 206

De la Rémy Moulin, Baie-Saint-Paul, Que. (c. 1726), 103

Desgagné Moulin à Vent, Ile-aux-Coudres, Que. (c. 1772), 24, 26

Dexter's Grist Mill, Sandwich, Cape Cod, Mass. (c. 1654), 103

Eastham Windmill, Eastham, Cape Cod, Mass. (c. 1688), 25, 26

Eleutherian Powder Mills, Hagley Museum, Wilmington, Del. (c. 1802), 120, 121, 122, 123

Elliot Grist Mill, Williamsford, Ont. (1870's), 4

Elora Flour Mill, Elora, Ont. (c. 1852), 84

Eplett's Flour Mill, Coldwater, Ont. (c. 1833), 12

Falls Grist Mill, Belvidere, Tenn. (1873), 116

Ferguson Grist Mill, Durham, Ont. (c. 1851), 87

Fowles Mill, Hastings, Ont. (c. 1850), 86

Franklin Cider Mill, Franklin, Conn. (c. 1890), 53

Frelighsburg Grist Mill, Frelighsburg, Que. (1839), 57

Furs' Feed Mill, Marchmount, Ont. (c. 1832), 78

George Washington Grist Mill, Mount Vernon, Va., (c. 1770; reconstructed 1930's), 86

Gilbert Stuart Grist and Snuff Mills, Saunderstown, R.I. (snuff mill c. 1753; grist mill c. 1757), 211

Glenora Mills, Glenora, Ont. (c. 1796), 126

Graue Mill Museum, Oak Brook, Ill. (c. 1852), 138

Grein's Feed Mill, Mount Forest, Ont. (c. 1866), 189, 192

Hagerman Grist Mill, Woodstock, N.B. (n.d.), 208

Haines Mill, Allentown, Penn. (1820's), 60, 83

Hapgood Carding Mill, Old Sturbridge Village, Mass. (1830's), 149

Harmony Textile Mills, Cohoes, N.Y. (c. 1867), 58, 142, 156, 160

Harris Mill, Harrisville, N.H. (1830), 132, 135

Harris Woollen Mill, Rockwood, Ont. (c. 1867), 204

Heckston Sawmill, Upper Canada Village, Morrisburg, Ont. (c. 1840), 113

Herr Grist Mill, Strasburg, Penn. (c. 1740), 164

Hope Sawmill, Keene, Ont. (c. 1836), 59, 187

Indian Mill, Upper Sandusky, Ohio (c. 1861), 82

Isaac Ludwig Mill, Grand Rapids, Ohio (c. 1846), 53

Ives Grist and Sawmills, North Tryon, P.E.I. (pre-1875), 83, 205

James Corwith Windmill, Watermill, Long Island, N.Y. (1800), 27

Jamestown Windmill, Jamestown, R.I. (1787), 2, 27

Jardin Zoologique de Québec Moulin à Vent, Orsainville, Que. (1932), 25

Juchereau-Duchesnay Seigneurial Mill, Saint-Roch-des-Aulnais, Que. (1700's), 103

Kintail, Mill of, Almonte, Ont. (c. 1830), 17

Krug's Sawmill, Chesley, Ont. (c. 1886), 112, 192, 193

Lang Grist Mill, Keene, Ont. (c. 1846), 53, 60, 61, 89

Leard Flour Mill, Coleman, P.E.I., (1888), 89

Legendre Mill, Stornaway, Que. (c. 1862; burned, rebuilt 1883), 190

L'Equille, Moulin de, Port Royal, N.S. (1607; reconstructed 1968), 66

Little River Sawmill, East Lebanon, Me. (c. 1774), 108, 112

MacAusland Woolen Mills, Bloomfield, P.E.I. (pre-1865; burned, rebuilt, 1949), 214, 215, 216, 217

MacLaren's Mill, Wakefield, Que. (1835; burned, rebuilt 1906), 188

MacPherson's Mill, Pictou Co., N.S. (c. 1861), 194

Mansur Sawmill, Weston, Vt. (n.d.), 59, 181

Marie Antoinette's Mill, Versailles, France (1780's), 52

Masse Lumber Company, East Vasselboro, Me. (c. 1797), 110, 114

McArthur Textile Mill, Carleton Place, Ont. (c. 1871), 1, 155

McClung's Grist Mill, Zenith, W. Va. (pre-1885), 190

McDonald Brothers' Sawmill, Sherbrooke Village, N.S. (c. 1826; reconstructed 1970's), 108, 113

Montarville, Moulin de, Mont-Saint-Bruno, Que. (c. 1710 in wood; rebuilt in stone 1741), 98

Moors Sawmill, Wilton, N.H. (1880's), 202

Morningstar Mountain Mill, Fonthill, Ont. (c. 1872; burned and rebuilt 1895), 85, 172

Nova Scotia Textiles, Windsor, N.S. (n.d.), 162

O'Hara Sawmill, Madoc, Ont. (1840's) 191

Old Batsto Grist Mill, N.J., (c. 1828), 83

Old Batsto Sawmill, Batsto, N.J. (c. 1882), 105, 112

Old Grist Mill (Frank Mansur Mill), Weston, Vt. (n.d.), 59, 181

Old Mill Museum, Youngstown, Ohio (c. 1845), 53

Old Red Mill, Clinton, N.J. (c. 1763), 82, 190

Old Slater Mill, Pawtucket, R.I. (1793), 147, 148, 149

Old Stone Mill, Delta, Ont. (c. 1795), 59, 184, 191

Papineauville Flour Mill, Que. (n.d.), 96

Pantigo Windmill, East Hampton, Long Island, N.Y. (c. 1771), 25, 55

Parent, Moulin de, Saint-Isidore, Que. (1810's), 109

Peirce Grist Mill, Washington, D.C. (c. 1820), 180

Pepperell Mills, Biddeford, Me. (c. 1848), 56, 156

Perkins Tidewater Grist Mill, Kennebunkport, Me. (c. 1749), 170, 180

Philipsburg Mill and Manor, North Tarrytown, N.Y. (c. 1720; reconstructed 1969), 92

Ponemah Textile Mills, Taftville, Conn. (c. 1871), 156, 159

Prescott Farm Windmill, Middletown, R.I. (c. 1811), 23, 25, 26, 30

Prescott Tower Windmill, Prescott, Ont. (c. 1822; now a lighthouse), 175

Robert E. Lee Mill, Stratford, Va. (c. 1755; reconstructed 1935), 14

Robertson's Corn Mill, Williamsburg, Va. (c. 1720; reconstructed 1957), 69

Roblin's Mill, Black Creek Pioneer Village, Toronto, Ont. (1842), 32, 58, 88, 89, 191, 193

Rosamond Textile Mill, Almonte, Ont. (c. 1866), 57, 58, 159

Saddle Rock Tidal Grist Mill, Great Neck, N.Y. (c. 1702), 61, 90, 190

Saint-Antoine-de-Tilly Carding Mill, Que. (n.d.), 59

Saugus Iron Works, Saugus, Mass. (c. 1646), 61, 70, 71, 72, 73

South Melville Grist Mill, South Melville, P.E.I. (n.d.), 201

Spocott Windmill, Lloyds, Md. (c. 1850; reconstructed 1971), 25, 27

Spring Mill, Mitchell, Ind. (c. 1817), 196

Steadman Grist Mill, Scituate, Mass. (c. 1640), 103

Stoney Brook Mill, Brewster, Cape Cod, Mass. (c. 1663; reconstructed 1893), 88, 190

Stony Brook Grist Mill, Stony Brook, N.Y. (c. 1750), 81

Thompson-Perkins Mill, Ottawa, Ont. (c. 1842), 171

Tom Walker's Grist Mill, Parshallville, Mich. (c. 1869), 219

Tonnancour, Moulin de, Pointe-du-Lac, Que. (c. 1721), 58, 60, 61

Tremblay, Moulin de, Les Eboulements, Que. (c. 1750), 58, 82

Trent Valley Woollen Mill, Campbellford, Ont. (c. 1881), 159

Union Mills, Union Mills, Md. (c. 1797), 199

Upper Canada Village (Heckston Sawmill), Morrisburg, Ont. (c. 1840), 113

Watson's Mill, Manotick, Ont. (1860), 53, 189

Wayside Inn Grist Mill, Sudbury, Mass. (c. 1702; reconstructed 1929), 86

Wight Grist Mill, Old Sturbridge Village, Mass. (reconstructed 1938), 211

Winthrop Grist Mill, New London, Conn. (1650's), 181

Wolf Pen Grist Mill, Louisville, Ky. (1830's). 64

The type faces used in this book are a film version of
Hermann Zapf's Palatino and Optima.
The typesetting is by Howarth and Smith Ltd.
The film separations are by Herzig Somerville Ltd.
The paper is The Warren Paper Company's 80 lb. Flokote
Sheets printed by Ashton-Potter Ltd.
Binding by T. H. Best Co. Ltd.

PRINTED AND BOUND IN CANADA